Y0-BTB-654

SIMON FRASER UNIVERSITY
W.A.C. BENNETT LIBRARY

UG 740 M5724 2005

MISSILE DEFENCE

This book examines the regional ramifications of the American plans for the deployment of a comprehensive missile defence system. As America's desire to secure its global leadership and influence is confronted by the insistence of other actors, such as Russia, China and NATO allies, to gain a prominent role in the international order, this volume analyses whether America's missile defence plans can satisfy these other powers and to what extent the US can ignore regional politics in its quest to design the global security architecture.

This book first provides an overview of the history of US missile defence efforts, relates this quest for defence to other strategic ideas such as offence and deterrence, and examines the likely motives behind the current US defence project. It then analyses the reactions and ambitions of major regional powers, notably Russia and China, and the dynamics of regions with less potent but still significant states, notably Europe and the Middle East.

This volume addresses these questions and breaks new ground by explicitly examining the global–regional tensions inherent in missile defence and the likelihood that the 'one vision' of the US will break into 'plural policies', shaped to meet regional conditions.

This book will be of interest to students and practitioners of strategic studies, security studies and international relations.

Bertel Heurlin is Jean Monnet Professor of European Security and Integration at the University of Copenhagen. He has written extensively on international politics, strategy, European security and integration, defence and security policy of the US, Russia, China, Germany and the Nordic states. He is a member of Danish defence commissions and of NATO scientific advisory committees.

Sten Rynning is associate professor at the Department of Political Science at the University of Southern Denmark. A former Fulbright Scholar and NATO Research Fellow, he has published extensively on NATO, the European Union and national security policy.

CONTEMPORARY SECURITY STUDIES

NATO's SECRET ARMY: OPERATION GLADIO AND TERRORISM IN WESTERN EUROPE
Daniele Ganser

THE US, NATO AND MILITARY BURDEN-SHARING
Peter Kent Forster and Stephen J. Cimbala

RUSSIAN GOVERNANCE IN THE TWENTY-FIRST CENTURY: GEO-STRATEGY, GEOPOLITICS AND NEW GOVERNANCE
Irina Isakova

THE FOREIGN OFFICE AND FINLAND 1938–1940: DIPLOMATIC SIDESHOW
Craig Gerrard

RETHINKING THE NATURE OF WAR
Edited by Isabelle Duyvesteyn and Jan Angstrom

PERCEPTION AND REALITY IN THE MODERN YUGOSLAV CONFLICT: MYTH, FALSEHOOD AND DECEIT 1991–1995
Brendan O'Shea

THE POLITICAL ECONOMY OF PEACEBUILDING IN POST-DAYTON BOSNIA
Tim Donais

THE RIFT BETWEEN AMERICA AND OLD EUROPE: THE DISTRACTED EAGLE
Peter H. Merkl

THE IRAQ WAR: EUROPEAN PERSPECTIVES ON POLITICS, STRATEGY, AND OPERATIONS
Edited by Jan Hallenberg and Håkan Karlsson

WEAPONS PROLIFERATION AND WAR IN THE GREATER MIDDLE EAST: STRATEGIC CONTEST
Richard L. Russell

PROPAGANDA, THE PRESS AND CONFLICT: THE GULF WAR AND KOSOVO
David R. Willcox

MISSILE DEFENCE: INTERNATIONAL, REGIONAL AND NATIONAL IMPLICATIONS
Edited by Bertel Heurlin and Sten Rynning

GLOBALISING JUSTICE FOR MASS ATROCITIES: A REVOLUTION IN ACCOUNTABILITY
Chandra Lekha Sriram

ETHNIC CONFLICT AND TERRORISM: THE ORIGINS AND DYNAMICS OF CIVIL WARS
Joseph L. Soeters

GLOBALISATION AND THE FUTURE OF TERRORISM: PATTERNS AND PREDICTIONS
Brynjar Lia

NUCLEAR WEAPONS AND STRATEGY: THE EVOLUTION OF AMERICAN NUCLEAR POLICY
Stephen J. Cimbala

NASSER AND THE MISSILE AGE IN THE MIDDLE EAST
Owen L. Sirrs

WAR AS RISK MANAGEMENT: STRATEGY AND CONFLICT IN AN AGE OF GLOBALISED RISKS
Yee-Kuang Heng

MILITARY NANOTECHNOLOGY: POTENTIAL APPLICATIONS AND PREVENTIVE ARMS CONTROL
Jurgen Altmann

NATO AND WEAPONS OF MASS DESTRUCTION: REGIONAL ALLIANCE, GLOBAL THREATS
Eric R. Terzuolo

MISSILE DEFENCE

International, Regional and National Implications

Edited by

Bertel Heurlin and Sten Rynning

LONDON AND NEW YORK

First published 2005
by Routledge
2 Park Square, Milton Park, Abingdon, Oxon OX14 4RN

Simultaneously published in the USA and Canada
by Routledge
270 Madison Ave, New York, NY 10016

Transferred to Digital Printing 2005

Routledge is an imprint of the Taylor & Francis Group

© 2005 Bertel Heurlin and Sten Rynning

Typeset in Great Britain by Keyword Group Ltd
Printed and bound in Great Britain by
TJI Digital, Padstow, Cornwall

All rights reserved. No part of this book may be reprinted or reproduced or utilised in any form or by any electronic, mechanical, or other means, now known or hereafter invented, including photocopying and recording, or in any information storage or retrieval system, without permission in writing from the publishers.

British Library Cataloguing in Publication Data
A catalogue record for this book is available from the British Library

Library of Congress Cataloging in Publication Data
Missile defence: international, regional, and national implications/
edited by Bertel Heurlin and Sten Rynning.
p. cm
ISBN 0-415-36120-6 (hardback)
1. Ballistic missile defenses. 2. National security. 3. Security, International. 4. World politics – 21st century. I. Heurlin, Bertel. II. Rynning, Sten, 1967–. III. Title.
UG740.M5724 2005
358.1′74 – dc22 2004028909

ISBN 0-415-36120-6

CONTENTS

FIGURES

CONTRIBUTORS

Wilhelm Agrell is Ph.D. and Associate Professor of History and is currently senior researcher at the Research Policy Institute, Lund University, Sweden. His main fields of research are Swedish and European defence and security policy as well as weapons developments in the Cold War period.

Tarja Cronberg is Dr.Tech. and Dr.Merc. specializing in the sociology of technology. She is currently a Senior Fellow at the Finnish Institute of International Affairs and a member of the Finnish Parliament. She is the former Director of COPRI, the Copenhagen Peace Research Institute, and Associate Professor at the Danish Technical University. Her latest publication is *Transforming Russia from Military to Peace Economy* (IB Taurus, London, 2003).

Ingemar Dörfer is Director of Research at the Swedish Defence Research Agency. He was Special Adviser to the Swedish Foreign Minister from 1992 to 1994 and has been Guest Scholar at the Woodrow Wilson International Center for Scholars, Brookings, SCIS, the Harvard Center of International Affairs, IISS, Stiftung Wissenschaft und Politik and the Norwegian Institute of International Affairs.

Peter Dyvad is a commander in the Royal Danish Navy and is currently working in the Danish Ministry of Defence

Birthe Hansen is Ph.D. and Associate Professor at the Department of Political Science, University of Copenhagen. She is also a foreign policy consultant, a lecturer at the Danish Defence Academy, and the director of a major research project on lost power and strategies of adaptation. She is the author of several books and articles on the Middle East, terrorism and US unipolarity.

Bertel Heurlin is Jean Monnet Professor in European Security and Integration at the Department of Political Science, University of Copenhagen. He is a member of Danish defence commissions and of NATO scientific advisory committees. His research focuses on American foreign, security and defence policy, European security, transatlantic relations, and the revolution in military affairs.

Alla Kassianova is Ph.D. and Associate Professor at the Department of International Relations, Tomsk State University. She has been visiting scholar at the Center for Political Studies, University of Michigan, Ann Arbor, and the Center for Nonproliferation Studies, Monterey Institute for International Studies. She has published in Russian and English on the foreign and security policy of the Russian Federation, including *Russian–European Cooperation on Theater Missile Defense*.

Kristian Søby Kristensen is assistant researcher at the Danish Institute for International Studies (DIIS). His current research is on the concept of war and the future conditions for the use of armed force.

Sten Rynning is Associate Professor at the Department of Political Science, the University of Southern Denmark. A former Fulbright Scholar and NATO Research Fellow, Rynning works on European and transatlantic security issues and has published extensively on NATO, the European Union and national security policy. His latest book is *NATO Renewed: The Power and Purpose of Transatlantic Cooperation* (Palgrave, New York, 2005).

ACKNOWLEDGEMENTS

This anthology is the second joint publication by 'The New Role of Military Forces', a research network within the framework of the Nordic Security Policy Research Program funded by the Nordic Council of Ministers and the Nordic Ministries of Defence. The authors met in Copenhagen under the auspices of the Danish Institute for International Studies (DIIS) to discuss the issue of missile defence and to organize this publication, and we should like to use this opportunity to convey our most sincere gratitude to the discussants who not only gave us their valuable time but also provided us with constructive comments. Our gratitude extends also to Terry Terriff, who served as our external reviewer subsequent to the authors' conference. He was not the only one to serve in this capacity: three anonymous reviewers of Frank Cass thoroughly reviewed the chapters and provided written comments. We are grateful for all the constructive input. Finally, our warm thanks to Stine Lehmann Larsen and Mia Lund Rasmussen on whose organizational and analytical skills the research network depended.

ABBREVIATIONS

ABM	Anti-Ballistic Missile (Treaty)
AMRAAM	Advanced Medium-Range Air-to-Air Missile
ASAT	Anti-Satellite System
ASW	Anti-Submarine Warfare
ATBM	Anti-Tactical Ballistic Missile
AWACS	Airborne Warning and Control System
BMC4I	Battle Management Command, Control, Communications, Computers and Intelligence
BMD	Ballistic Missile Defence
BMDS	Ballistic Missile Defence System
BMEW	Ballistic Missile Early Warning
BW	Biological Weapons
C4ISR	Command, Control, Communications, Computers, Intelligence, Surveillance and Reconnaissance
CFE	Conventional Forces in Europe
CIA	Central Intelligence Agency
CIS	Commonwealth of Independent States
CSTO	Collective Security Treaty Organization
CTBT	Comprehensive Test Ban Treaty
CW	Chemical Weapons
CWC	Chemical Weapons Convention
DEW	Distant Early Warning
DoD	Department of Defense (US)
DOTMLPF	Doctrine, Organization, Training, Material, Leadership, Personnel and Facilities
EADS	European Aeronautic Defence and Space (Company)
ESDP	European Security and Defence Policy
EW	Early Warning
EWR	Early Warning Radar
FMCT	Fissile Material Cut-off Treaty
GBI	Ground-Based Interceptors
GPALS	Global Protection Against Limited Strikes

HEO	Highly Elliptical Orbit
IAEA	International Atomic Energy Agency
ICBM	Intercontinental Ballistic Missile
IDC	Initial Defensive Capability
INF	Intermediate Nuclear Forces
IRBM	Intermediate-Range Ballistic Missile
ITU	International Telecommunication Union
MAD	Mutual Assured Destruction
MD	Missile Defence
MEADS	Medium Extended Air Defence System
MFA	Ministry of Foreign Affairs (Russia)
MILCON	Military Construction
MIRV	Multiple Independently Targetable Re-entry Vehicles
MRBM	Medium-Range Ballistic Missile
MRV	Multiple Re-entry Vehicles
MTCR	Missile Technology Control Regime
NATO	North Atlantic Treaty Organization
NBC	Nuclear, Biological and Chemical
NCA	National Command Authority
NIE	National Intelligence Estimate
NMD	National Missile Defence
NORAD	North American Aerospace Defense Command
NPT	Nuclear Non-Proliferation Treaty
OSCE	Organization for Security and Cooperation in Europe
OTH	Over the Horizon
R&D	Research and Development
RAF	Royal Air Force
RAMOS	Russian–American Observation Satellite
RDT&E	Research, Development, Test and Evaluation
RMA	Revolution in Military Affairs
SAAM	Surface-to-Air Anti-Missiles
SAC	Strategic Air Command
SALT	Strategic Arms Limitation Talks
SDI	Strategic Defence Initiative
SIPRI	Stockholm International Peace Research Institute
SLBM	Submarine-Launched Ballistic Missiles
SRF	Strategic Rocket Forces
SSBN	Nuclear-Powered Ballistic Missile Submarine
START	Strategic Arms Reduction Talks
STRATCOM	United States Strategic Command
TMD	Theatre Missile Defence
WMD	Weapons of Mass Destruction

INTRODUCTION

Bertel Heurlin, Sten Rynning and Kristian Søby Kristensen

In twelfth-century Latin Christian Europe, the demise of knights inaugurated a new and more complex art of war based on the interaction of crossbows, pikemen and flanking cavalry. In the new era, war demanded not merely headlong charges by noblemen who had made it their business to rule and make war but also organization, training and coordination. Some new political units were able to benefit, many old ones were not; in consequence, the demands of war 'reinforced localism' and 'led to the collapse of the imperial fabric'.[1]

This tension between localism and the imperial fabric is an often overlooked aspect of the US missile defence to which President George W. Bush has committed the United States.[2] The US is a modern empire whose military forces sustain its conception of order and preferred way of life. Like other empires before it, the US must develop military forces that reinforce its strategic position and keep it abreast of local disputes.[3] Maybe missile defence will allow it to do so and thus avoid the fate of the Holy Roman Empire and papacy in the twelfth century, but the issue should be an object of study rather than faith.

The fact that President Bush has committed the US to deployment allows us to focus more clearly on the regional implications of America's missile defence. Previously, the lack of a commitment made scholars and observers gravitate towards the centre of decision-making, Washington DC and Beltway politics, in anticipation of an insight into the decisive dynamic that would stack the cards in favour of missile defence proponents or opponents. Many insights were thus gained, but studies tended nevertheless to reflect the ideological and speculative nature of the missile plans themselves. We now know that the effort will be made to erect a defensive system, and we can thus begin to focus more sharply on the interaction between this US design for global order and regional questions of cooperation and conflict.

In consequence, this book is located at the crossroads of US security policy and regional security dynamics, acknowledging US pre-eminence and the power of the US to shape outcomes but likewise the power of regional dynamics, especially in the post-Cold War era, to inhibit US policy. The book stands on the shoulders of the many other books and articles that have examined the making of Bush's missile defence commitment, but it also seeks to fill in a gap by putting the issue in a wider, international context.

The remainder of this introduction is divided into three parts. The first part reviews the history of missile defence and thus the many decision-making phases that have resulted in the current deployment commitment. The second part engages the missile defence literature and outlines the argument that the real political potential of missile defence must be appreciated at the intersection of US policy and regional dynamics. The final section outlines the individual contributions of the book and also the themes that will be examined again in the conclusion.

The history of missile defence: MAD v. rogue states

Missile defence is not new in the history of international relations. Crossbow bolts were among the first missiles; as a defence, the Second Lateran Council of 1139 simply banned their use among Christians. As missiles have become more potent, defensive means have evolved, at least in their conception. The most potent missiles have been and continue to be intercontinental ballistic missiles (ICBMs) armed with weapons of mass destruction (WMD). As ICBMs began to be developed from the 1950s and onwards, the Cold War protagonists searched for defensive measures but also failed to develop them. The state of technological development at the time accounts for part of this failure, but the failure can also be traced to mass: it was simpler to build offensive arsenals and therefore simple to overwhelm defensive systems with huge offensive arsenals. Defence and the associated idea that the adversary's goals can be denied were therefore replaced by strategies of deterrence based on offensive capabilities and the threat of mutual assured destruction (MAD).

In 1972, the theory of deterrence was institutionalized in the Anti-Ballistic Missile (ABM) treaty, which placed strict limits on the superpowers' ability to develop and deploy missile defences. Simultaneously, the superpowers sought to contain the offensive arms race by placing caps on their arsenals – leading to strategic arms limitation agreements (SALT) in 1972 and 1979.

The embrace of mutual vulnerability was both tentative and disputed, and various actors continued to exert pressure for effective defences. The Soviet Union worked within the confines of the ABM treaty to defend its capital, Moscow. In the US, in the 1960s, key political actors were beginning to accept the idea of vulnerability in the Soviet case but not in the Chinese case, which led to the defence project Sentinel.[4] President Nixon, elected in 1968, abandoned the system in favour of a new system, Safeguard, in line with the doctrine of deterrence because the system was designed merely to protect the offensive weapons of the US. When Safeguard was abandoned due to technical difficulties it might have seemed that the US was ready to accept the deterrence doctrine as its guiding strategy. This was certainly the case for a number of liberal groups but not for conservatives, who channelled their frustrations into the Committee on the Present Danger and thereby helped set the agenda that Ronald Reagan propagated once elected in 1981.[5]

In his famous so-called 'Star Wars' speech in 1983, President Reagan questioned both the rationality and the morality of the strategic policy of deterrence. In Reagan's view it was immoral to leave the American population vulnerable to Soviet missiles and he found that 'truly lasting stability' would depend on effective defence, not deterrence.[6] President Reagan envisioned a policy that would render 'nuclear weapons impotent and obsolete' and he proposed the Strategic Defence Initiative (SDI) to realize it.[7]

Reagan engaged in 'a technological endgame'[8] with the purpose of terminating the competition between the two superpowers. Predictably, the announcement set off a new defensive arms race because the Soviet Union was faced with a simple choice – to follow suit or to give in. Soviet leaders may have been inclined to compete but they lacked the technological and economic means and were therefore forced to negotiate the end of the Cold War and, effectively, the Soviet empire. Reagan's SDI had thus achieved a victory while still at the stage of declaratory policy and scientific research. This victory, paradoxically, also undermined the programme because its geopolitical justification vanished.

During the early 1990s, missile defence receded to become just one of many DoD research and development projects. To be sure, the Gulf War in 1991 had shown that the US in the post-Cold War environment could face a threat from ballistic missiles, but the only powers with an ability to strike at the US homeland were the 'old' nuclear powers – Britain, France, Russia and China – and not the regional strong-men whose strategic rationality was questioned in the wake of Saddam Hussein's defeat. Prodded by this debate and also by the apparent success of the Patriot system during the 1991 war, US research focused on theatre missile defence and the ability to protect deployed troops.[9]

The year 1998 turned out to be a decisive turning point. The bipartisan Rumsfeld Committee tasked to evaluate the importance of missile threats reached a conclusion that found wide backing in the Republican-dominated US Congress: the threat from ICBMs was not a phantom of the Cold War but a stark reality of post-Cold War geopolitics. The Rumsfeld Committee thus gave birth to a new intense debate on how to fashion a missile defence appropriate for the new age.[10]

The Rumsfeld Committee picked up the theme of strategic rationality that had emerged during the 1991 Gulf War and essentially made the case for missile defence by identifying a new type of threat. Thus, the existing stockpiles of nuclear-armed ballistic missiles of the old nuclear powers are not a threat. Instead, rogue states are. Rogue states are small, under-developed countries with ambitions to revise regional geopolitics and whose designs for revision are threatened by the American power to intervene and preserve the status quo. Rogue states are therefore incited to acquire ballistic missiles armed with various WMD and constrain American policy by holding a segment of the American population – a large city, for instance – hostage. From an American perspective, there is no doubt that regional interventions occasionally are necessary to maintain order, and the question of whether the 'imperial fabric' can withstand rogue nations' challenges is therefore of essence. Two particular aspects make missile defence a viable and necessary option in the face of these new threats, according to the Rumsfeld Commission. First, rogue states are irrational in the sense that they cannot be trusted to act in the interest of their people, and the threat of massive retaliation may therefore be useless, thus invalidating the theory of deterrence. A new strategic policy is therefore required. Moreover, the capabilities of these actors are limited, and while even limited WMD capabilities can inflict tremendous damage, this does widen the range of possible defensive measures.

It is important to note that virtually all aspects of the Rumsfeld Committee's conclusions and the missile defence plans that followed have been disputed. Some observers questioned the assumption that rogue states could not be deterred;[11] others the technological feasibility of the system;[12] and political actors within the US and outside of it voiced concerns about the impact of strategic defence on the global order and notably the danger that the end to the doctrine of mutual vulnerability would translate into the beginning of a new arms race and thus also strategic instability.[13]

Missile defence thus cannot be reduced to the simple equation of American status quo and rogue revision: the issue has many political, economic and technological dimensions that are distinct but also interrelated. The complexity of the issue helps explain why the Clinton presidency was not entirely committed to the issue of missile defence, although President Clinton turned out to support the idea in principle. First, the system had yet to convincingly prove its technological feasibility, which questioned both the military and economic rationality of proceeding towards actual deployment. Second, the European allies along with China and Russia repeatedly made clear their opposition to missile defence and thus effectively promised to accompany any deployment with strained international relations. As we know, President Clinton considered the political costs and technological uncertainties too high and therefore postponed the deployment decision beyond his presidency.

George W. Bush came to office with a markedly more positive view on missile defence and did not hesitate to follow up on his campaign promises to place the issue on top of the national security agenda. President Bush was initially faced with exactly the same problems as President Clinton, although that did not prevent his administration from pursuing the issue on all fronts. The problems diminished in the wake of the September 2001 terrorist attacks, however, as the scope for unilateral US action increased. Thus, the political costs of deciding to deploy were substantially lowered and in December 2001 George W. Bush announced, principally to Russia but naturally to all others as well, the unilateral withdrawal of the US from the ABM treaty. Unconstrained by international agreements, the Bush administration embarked upon the deployment of a functional missile defence system.

Currently the United States is deploying an embryonic missile defence capability with the stated purpose of defending the US homeland against a limited ballistic missile attack. This actual deployment is taking place simultaneously with the further testing of a range of different subsystems that will be incrementally integrated in the existing system, in time creating a layered missile defence system capable of protecting the whole of the US homeland as well as deployed troops and friends and allies against a limited missile attack.

The question of 'whether' to deploy a missile defence is a question of the past; missile defence is rapidly becoming a fact. It is therefore appropriate to take stock of the international consequences of the project and answer some of the questions that the pre-deployment debate either left out or was unable to answer.

The argument of the book: unipolarity and regional dynamics

The American argument for the deployment of a missile defence system presented in the previous section is shaped by the realities of the post-Cold War international system and notably the fact that the US is the only remaining superpower – a unipole. Decisions made in Washington have global ramifications and many observers naturally focus on the US capital when assessing the impact of current trends. This is certainly the case for the literature on missile defence, which to a large extent is policy-oriented and thus presents strategies of action and policy proposals to American decision-makers: discussing what configuration of missile defence might bring with it least harmful international implications,[14] whether it would enhance American security,[15] or if the arguments presented by the administration stand up to closer scrutiny.[16] These studies may be useful for American decision-makers but they also downplay or even overlook wider political consequences of American missile defence deployment and in particular the regional dynamics it could set in motion. The issue is too often defined in terms of a narrowly defined threat/response relationship – involving the US and rogue nations – and this book is part of the effort to bring a new dimension into the debate.

The book naturally recognizes the unique unipolar position of the United States and the fact that regional dynamics will be significantly influenced by US policy. Still – and this is the book's basic *raison d'être* – unipolarity is not omnipotence. The US might be able to shape regional dynamics, but these dynamics will over time also contribute to the shaping of US policy.

Avoiding or circumventing the blind spots inherent in the often implicit unipolar perspective is appealing but also raises the difficult question of how to analyse a diverse global chessboard. Diversity has left its mark on the record of responses to the current missile defence blueprint: different regions and actors react in different ways. There is little to do but to engage this record on its own terms and trace reactions, their sources and motivations, and their regional and international impact.

The new diversity of the international system is reflected in the keywords that mark the literature on international relations – such as multipolarity, interdependence, pluralism, or the global/local nexus – and it reflects a widespread concern with the ramifications of change in the international structure. This book is part of this trend in so far as it presupposes that the ability of local actors to envision a particular political future for themselves and their environment and to act upon these visions has increased. We are not claiming to break new ground here: other people have already examined the relation between a particular region and missile defence.[17] We do stake a claim to some novelty, however, in so far as the current literature predominantly focuses on a particular region – and often either Europe or Asia – and does not seek a more global view of regionalization.[18]

The book seeks to balance the global and the regional. The global perspective is centred on the US – its ambitions and defence plans – and is important partly because of the structural power of the US, partly because it checks the temptation to go native in the assessment of particular regions. The regional perspective, as already mentioned, is critical in the new, emerging international system.

In sum, we aspire to work at the crossroads of US security policy and regional dynamics and present a broad but also analytically focused presentation of the American missile defence plans and their wide-ranging international implications. Individual chapters in the book will add regional insight; the collective work will provide global comprehension.

The structure of the book

The most fundamental question in need of an answer in relation to missile defence is why? Why does the American administration continue to follow its quest for missile defence even when many commentators point to the fact that terrorism in general is a much more pressing threat, and, even if terrorists or their state sponsors were in fact to use WMD against the US, why then pursue missile defence when ballistic missiles would arguably not be their preferred means of delivery?[19] The threat from rogue states cannot single-handedly explain this critical question.

Accordingly, the different rationalities driving the American missile defence plans are the subject of the three first chapters. Each of these chapters presents us with different but complementary arguments enhancing our understanding of current missile defence policies. In the first chapter, Wilhelm Agrell discusses the historical development of air defence strategies in general and focuses especially on the differences between the present situation and the Cold War. The most striking difference is that whereas defensive measures during the Cold War rested on two essentially opposing strategies, the arms race and arms control, they still had as an overall goal the maintenance of strategic stability. This goal has changed. Beginning with the SDI project, stability is no longer the main goal of the United States. As stated by Wilhelm Agrell in his chapter: 'In a unipolar world, the remaining hegemonic superpower has no need for deterrence through balance, on the contrary.' The new international situation opens up a range of alternative strategies to follow for the US, where deterrence is only one among others. Instead of being limited to the strategy of balance, options such as pre-emption, dissuasion and interception are all viable paths to increased national security. In fact, missile defence is a critical component to various combinations of these strategies, and Agrell's chapter helps us understand this context of evolving strategic thinking.

Tarja Cronberg likewise proceeds from the unipolar position of the US, but her argument relates not so much to the relationship between the unipole and its contenders as to the wish of the unipole to remain just that. In short, technology as much as rogue states is driving missile defence. Continued technological superiority is a potent way to maintain unipolarity, and missile defence is thus not an end in itself; it is part of a larger strategy – a strategy of dissuasion through technology. The most critical dimension of missile defence in this context is found in outer space, which attracts both military and civilian investments in the US. Missile defence and space have two interfaces. On the one hand, missile defence as a research project increases the American technological capacity to operate in space writ large. Moreover, the deployment of a comprehensive missile defence will result in a large increase in the capabilities needed to protect both civilian and military space-based infrastructure. Accordingly, in Cronberg's analysis, missile defence can only be understood if one combines an assessment of technology and space policy with that of rogue states.

In the third chapter, Bertel Heurlin investigates the war and peace rationales inherent in the US drive to establish a missile defence. Heurlin raises a question at the outset of his analysis, namely, why a symmetrical means – missile defence – is capturing so much attention in an era of asymmetrical warfare, witnessed by terrorism and insurgency. Heurlin finds the answer in distinct dimensions of US military strategy, which, as emphasized in the preceding chapters also, evolves around the pre-eminent question of sustaining unipolarity. First, Heurlin notes that the US is actively redefining the attributes of being a superpower. In the past, superpowers were defined in part by their nuclear arsenals; today, and even more so in the future, superpowers are 'anti-nuclear' in the sense that they are capable of denying others the strategic fruits of an offensive nuclear capability. Moreover, missile defence ties in with the heralded 'revolution in military affairs', which is based on constant transformation and innovation. Missile defence research is, then, one of several research projects that will exhaust contenders before the competition has really begun. Finally, Heurlin concurs with Tarja Cronberg's argument that outer space is an important new dimension in US strategic thinking.

These three chapters deal with the familiar theme of unipolarity, but they all underscore that missile defence is not solely a question of responding to rogue threats. In its efforts to sustain the unipolar moment, the US is looking for means to preserve and possibly extend its strategic freedom of action. Missile defence is one component: it not only secures freedom of action during regional interventions where the US may encounter rogue nations armed with missiles and WMD, it also enables new combinations of strategic policy and adds a technological dimension to US superiority, particularly in outer space.

The remainder of the book deals with the regional ramifications of this policy. We start in Chapter 4 with an analysis of Russian responses to the American missile defence plans. During the 1990s, a major concern of many observers was the Russian reaction to an American withdrawal from the ABM treaty: the fear of an extreme response and concomitant strategic instability was widespread. Surprisingly, and fortunately one might add, the Russian response was rather low-key and the major diplomatic confrontation did not take place. Instead, the two former adversaries have engaged in discussions concerning the possibilities of missile defence cooperation. That is the point of departure for Alla Kassianova, who notes that two incompatible positions exist on the political scene in Russia, a cooperative and a competitive. Kassianova examines developments within Russia as well as internationally in order to assess the likely winning position in Russia. She concludes that the potential for cooperation is limited. To be sure, consultations between Russia and NATO are ongoing, but it seems that political and economic actors are more in support of continuing the indigenous Russian missile defence programs that were boosted by the US decision to withdraw from the ABM treaty, which will likely lead to the deployment of various new countermeasures in the Russian strategic forces.

European allies follow American missile defence plans closely owing to the obvious impact of these plans on the transatlantic alliance and the coupling of transatlantic security agendas. The European allies have not been overtly positive towards the concept of missile defence because the logic of deterrence and mutually assured destruction in the eyes of many Europeans is adequate for maintaining stability. In his chapter on the European region, Sten Rynning points out that divergent threat assessments account for the transatlantic missile defence disputes that characterized particularly the latter half of the 1990s. He also reaches the conclusion, however, that the European allies increasingly support the current deployment plans and he stakes the conclusion that this trend will continue. A primary reason for this turn of events is the behaviour of regional renegades, or rogue nations, whose efforts to secure WMD and missiles have become more marked in recent years, and whose political significance has changed post-September 2001. Europeans continue to define threats more broadly and thus also to advocate broader – political and economic in addition to military – strategies for countering them. But on the issue of missile defence, the convergence across the Atlantic has reached a point where pragmatics – the question of gaining defence investments and influence on command and control systems – will make an impact in favour of cooperation.

The third case study moves to Asia and focuses on the emerging great power, China, and the regional ramifications of its likely reactions to the American missile defence plans. China has few reasons to approve of these plans, Peter Dyvad argues. China's nuclear deterrent is limited in size and its political value will soon be undermined by even the limited US missile defence shield unless China makes significant new investments in its strategic nuclear forces. Thus, it will demand time and money for China simply to maintain the nuclear status quo, i.e. deterrence. Still, even though China otherwise is focused on economic and social modernization, it will be geopolitically motivated to invest also in nuclear modernization. Chinese leaders perceive a new type of containment behind the US plans and, significantly, a containment of China; for instance, the US missile defence system includes 'tactical' components that will shield notably Taiwan but also other regional allies, making it distinctively more difficult for China to obtain diplomatic gains through shows of military force. One might fear that China's response will be to ignite regional fires – for instance by increasing pressure on Taiwan, blocking US efforts to reach a nuclear weapons agreement with North Korea, or exporting more missile technology to Pakistan. Yet, most of these measures could in the end hurt China itself, Peter Dyvad concludes, and China will in the short to medium run likely focus on the modernization of its strategic forces while working to preserve stability on its borders. However, strategic competition between China and the US will continue, and regional stability in the long term does not follow automatically from the benevolent short-term perspective outlined here, the author warns.

Moving from the Far East to the Middle East, Birthe Hansen explores the region with perhaps the greatest potential for confrontation and war. The region has witnessed the use of weapons of mass destruction on several occasions and is infested by numerous unresolved disputes and rivalries, including hot confrontations in several locations. Still, the region's descent into greater violence may not take place, and missile defence may be one of the pieces in the political puzzle preventing it from doing so. Missile defence is a vital component in the American armour and thus in the US strategy of dissuasion: regional revisionists know that the US may not be deterred even if they possess a nuclear capability, and with missile defence the costs of building a nuclear deterrent may in themselves be dissuasive. This is capital for positive change, Birthe Hansen argues, but it must be spent as part of a greater investment package that includes notably assistance to democratizing regimes and support for arms control regimes. It follows that instability is a constant threat that will rise if arms initiatives such as missile defence are not wrapped in comprehensive policy packages.

Turning to a much smaller and more peaceful region, the chapter of Ingemar Dörfer focuses on Scandinavian reactions to missile defence. Scandinavia is part of Europe and therefore overlaps with the focus of Sten Rynning's contribution; still, Scandinavia is an interesting case of a sub-region in which countries feel unaffected by the new threat from rogue nations but still give up their adherence to Cold War concepts of stability at distinct rates. Ingemar Dörfer maps out the extent to which Scandinavia became a particularly peaceful region following the end of the Cold War but, as Dörfer argues, this did not imply that policy-makers should shy away from adapting policy concepts to new international realities. Dörfer then traces the way in which change has varied – Norway and Denmark have proved more adaptable than Sweden and Finland – and he traces the domestic political reasons why this is so.

The final chapter by Kristian Søby Kristensen engages in a subject of particular relevance to the global design on the US defence plans: the need to acquire overseas bases. In principle, the effort to secure such bases involves the US and a local government, both of whom negotiate the terms of reaching a security policy agreement. In practice, the effort involves a host of local and domestic issues in addition to the external security imperative. Kristensen demonstrates the way in which the effort is thereby made more complex. For instance, the American request for a new agreement in Greenland sparked renewed efforts on the part of the Greenland authority to secure greater independence from Denmark. However, Kristensen also presents the way in which the US government can use such a complex issue to further its own security policy agenda, by linking issues, pausing negotiations, and providing concessions in related domains. The complexity of the new world, Kristensen tells us, can be a diplomatic resource for the US if the complexities are well understood and then handled with both good fortune and skill.

The Conclusion will summarize findings and suggest new venues of research, but a foregone conclusion is that missile defence is becoming a reality and that the US agenda will impact differently on different regions of the world. This book cannot answer all questions related to missile defence, regionalism and world order, but it hopes to contribute to the research agenda and also to provide answers to the questions that found their way into these pages.

Notes

1 William H. McNeill, *The Pursuit of Power*, Chicago: University of Chicago Press, 1984, p. 68.
2 President George W. Bush, 'President Announces Progress in Missile Defense Capabilities', Statement by the President, The White House, 17 December 2002. Online. Available HTTP: ⟨http://whitehouse.gov/news/releases/2002/12/20021217.html⟩.
3 Barry Posen, 'Command of the Commons: The Military Foundations of US Hegemony', *International Security*, 28(1), Summer 2003, pp. 5–46.
4 The US had earlier attempted to build another defensive system, Nike Zeus, which involved the detonation of nuclear warheads to intercept incoming ICBMs. The programme went through various phases and was renamed in several instances, resulting in the decision under President Johnson in the 1960s to develop only a limited defence capability – Sentinel – to counter a smaller power like China and not the Soviet Union.
5 Roger Handberg examines this and the structure of other missile defence debates within the US from the 1950s to the present in *Ballistic Missile Defense and the Future of American Security: Agendas, Perceptions, Technology, and Policy*, New York: Praeger, 2001.
6 President Reagan, 'Address to the Nation on Defense and National Security', 23 March 1983. Online. Available HTTP: ⟨http://www.reagan.utexas.edu/resource/speeches/1983/32383d.htm⟩.
7 For an analysis of the impact of SDI on the Cold War, see Mira Duric, *The Strategic Defence Initiative: US Policy and the Soviet Union*, London: Ashgate, 2003; for an assessment of Ronald Reagan's intellectual background and susceptibility to accept a grand defence initiative, see Francis Fitzgerald, *Way Out There in the Blue: Reagan, Star Wars, and the End of the Cold War*, New York: Simon & Schuster, 2000.
8 See Wilhelm Agrell's contribution to the present volume.
9 The actual efficiency of the Patriot system has been widely contested, however; see especially Theodore A. Postol, 'Lessons of the Gulf War Experience with Patriot', *International Security*, 16, Winter 1991–2, pp. 119–171.
10 See for instance Bradley Graham, *Hit to Kill: The New Battle over Shielding America from Missile Attack*, New York: Public Affairs, 2001. The executive summary of the Commission report is available online. Available HTTP: ⟨http://www.fas.org/irp/threat/bm-threat.htm⟩.

11 Steven E. Miller, 'The Flawed Case for Missile Defense', *Survival*, 43, 2001, pp. 97–98.
12 Richard L. Gavin, 'A Defense That Will Not Defend', *The Washington Quarterly*, 23, 2000, pp.116–118; Craig Eisendrath, Melvin A. Goodman and Gerald E. Marsh, *The Phantom Defense: America's Pursuit of the Star Wars Illusion*, New York: Praeger, 2001.
13 See for instance Michael Krepon, *Cooperative Threat Reduction, Missile Defense, and the Nuclear Future*, Basingstoke: Palgrave, 2002; see also Handberg, op. cit., and Graham, op. cit. for the domestic US debate.
14 Ivo H. Daalder *et al.*, 'Deploying NMD: Not Whether, But How', *Survival*, 42, 2000, pp. 6–42.
15 Clifford Singer, 'How Can National Missile Defense Best Enhance Security', *Defense and Security Analysis*, 18, 2002, pp. 293–301.
16 Miller, op. cit., pp. 95–110.
17 The literature on specific regions is extensive, but for the transatlantic relationship see Wyn Q. Bowen, 'Missile Defence and the Transatlantic Security Relationship', *International Affairs*, 3, 2001, pp. 485–507; and Philip H. Gordon, 'Bush, Missile Defense and the Atlantic Alliance', *Survival*, 1, 2001, pp. 17–36. For the consequences of missile defence in Asia, see Jianxiang Bi, 'Uncertain Courses: Theater Missile Defense and Cross-Strait Competition', *Journal of Strategic Studies*, 3, 2002, pp. 109–160; and Richard L. Russel, 'Swords and Shields: Ballistic Missile Defenses in the Middle East and South Asia', *Orbis*, 3, 2002, pp. 483–498. For the Middle East, see Arieh Stav, *The Threat of Ballistic Missiles in the Middle East: Active Defense and Counter-measures*, Brighton: Sussex Academic Press, 2004.
18 A notable exception is the volume edited by James Wirtz and Jeffrey A. Larsen, *Rockets' Red Glare: Missile Defenses and the Future of World Politics*, Boulder, Colo.: Westview Press, 2001. The book examines the ABM regime and arms control questions but also, in the third section, regional implications (although the chapter on 'allies' is thematic rather than regional). Other books tend to treat regional impacts in one chapter in an analysis often primarily focused on US policy: see for instance James E. Lindsay and Michael E. O'Hanlon, *Defending America: The Case for Limited National Missile Defense*, New York: Brookings, 2001, ch. 5.
19 Dean A. Wilkening, 'Keeping National Missile Defense in Perspective', *Issues in Science and Technology*, Winter 2001, pp. 51–52.

1

PRE-EMPT, BALANCE OR INTERCEPT?

The evolution of strategies for coping with the threat from weapons of mass destruction

Wilhelm Agrell

The nature of defensive choices

From where did the concept of missile defence emanate? And why has missile defence been and continues to be a controversial issue and not, as so many other inventions in the continuous chain of measures and countermeasures, just another self-evident instrument in a defence posture? The purpose of this chapter is to look more closely at these questions and search for the specific aspects of missile defence not so much as a technological system but as a strategic option. Missile defence is here discussed in the broader context of active air defence from the world wars to the present. The evolution of strategies of missile defence is a predominately, but not exclusive, US phenomenon. The Soviet and Russian experience is dealt with in another chapter.

My point of departure is a general concept or model for response to new strategic offensive systems. The interaction of measures and countermeasures over the various fields of warfare basically reflects similar strategic trade-offs and choices. Which are the defensive or responsive options available? And why do states and military organizations choose among them the way they do?

In the great battleship arms race preceding the First World War, the obvious choice for the major contenders was to construct bigger, more heavily armoured and armed and faster battleships at a rate matching or outbuilding potential opponents. For the secondary naval powers this was not within reach and other options had to be considered. One was to adjust overall naval ambitions and concentrate on coastal defence. The purpose of

Abolition

Non-proliferation

Deterrence

Pre-emption

Counterforce

Forward defence

Area defence

Point defence

Protection

Suppression

Figure 1.1 Possible counter-strategies for coping with weapons of mass destruction.

coastal defence battleships was not to match the Dreadnought-type battleships but to force a major naval power to commit the most powerful ships in an engagement with a secondary navy.

But as new technologies emerged, new strategic choices became available. Instead of countering gunships with gunships, *La Jeune École* in the late nineteenth century emphasized technological change, countermeasures and alternative strategies. The giants should not be met with giants but with new strategies based on smaller ships and torpedoes, successfully employed in the First World War by the Italians against the Austro-Hungarian navy. And in the Second World War another technological system, ground- and carrier-based aircraft, emerged as both the ultimate countermeasure and a superior replacement for the battleships.

Nuclear weapons forced all the major powers to consider possible counter-strategies for coping with the specific kind of threat posed by weapons of mass destruction. These emerging counter-strategies can be described as a number of choices along a ladder or a stair, where the highest steps represent the forward and preventive strategies and the lowest the most passive and reactive strategies (see Figure 1.1).

Abolition is an ultimate strategy; a non-existing (or preferably non-invented) threat does not require any further counter-strategies. Non-proliferation is a choice in a situation where not the weapons themselves but rather their diffusion is perceived as the main problem, as was the case after deterrence was established between the superpowers. Without deterrence there is a range of operational options for preventing, blunting, fending off or limiting a strike with WMD. This can be done by attacking production facilities, bases and offensive weapons systems, or through defensive measures through successive layers. Finally, passive measures can be employed with the aim of damage limitation. And of course, in the best of worlds the population can be reassured that no such thing as a devastating attack with WMD will ever take place.

These reactive/pre-emptive choices have three main dimensions. The first is about technology and resources. Civil defence became a key strategy in early Cold War planning since the air (and later missile) defence problem could not be solved. And civil defence was largely abandoned as the destructive power of the nuclear arsenals surpassed any protective countermeasure. The second dimension is about warfare and force postures. The shift from massive retaliation to flexible response in the 1960s was an attempt to redefine the role of nuclear weapons in warfare and thus to reduce the need for reactive defensive measures. The third dimension is about international politics. The growing number of threshold states in the 1960s compelled the international community, with some important exemptions, to support the establishment of the Non-Proliferation Treaty, thus transforming proliferation issues from bilateral relations to a multilateral international regime. This dynamic of technology, warfare and international politics remains at the core of the issue of strategies to counter WMD threats.[1]

It might be tempting to regard the 'ladder' above as a menu of choices ranging from less demanding to more demanding or from worse towards better. This is not entirely the case; as with naval warfare in the early twentieth century, strategic choices are more complicated than that, depending on whose choice it is and what political and technological circumstances are at hand. Non-proliferation, the most ambitious preventive strategy short of complete abolition of WMD, is since the 1960s the main strategy pursued by most small and medium powers, if for no other reason than the simple lack of alternatives for protection against the effects of nuclear war. And as several air campaigns have illustrated, from Korea and Vietnam to the Kosovo war, concealment, dispersion and sheltering can under specific circumstances be an effective countermeasure, reducing or even in extreme cases nullifying the effects of the offensive air operations employing conventional weapons.

The main element is, however, the specific nature of nuclear (and to a lesser extent chemical and biological) weapons. The gradual realization of the consequences of nuclear attack has as a consequence separated a WMD threat from that of long-range offensive weapons systems in general.

In the world of strategic abstraction, every state would of course at all times choose the solution or combination of solutions that gives the best and most credible protection. There is no sense, in this ideal world, in picking the second or third best, and certainly not for retreating down the ladder. However, in the real world, with all operational, technical, financial, political and psychological constraints that constitute the context of decision-making and strategic choices, things of course look quite different. It is empty at the top because credible abolition is not feasible and non-proliferation at best is a leaking protective wall. Weapons of mass destruction are based on technologies that cannot be bottled up and thus will remain as a real or potential threat.

A number of small and medium-sized countries had to back off from the option of national nuclear deterrence, either for lack of financial, technical or natural resources or for strategic and political reasons. How much is it reasonable to spend on exclusive weapons systems that, when the moment comes, you are unable to employ? And to what extent do the benefits of a weapons programme outweigh the prize in terms of sanctions, blacklisting and withdrawn benefits?

Technology is a variable, not a constant, but at any given moment it constitutes a limiting factor. The RAF and the British anti-aircraft defence in 1944 could not intercept incoming V-2 missiles for the simple reason that no weapons system with such capability existed and no technology for the construction of one was at hand. So, if not completely ignoring the threat by retreating down the ladder, a choice had to be made between possible existing options.

Suppression could in some sense represent the ultimate defensive strategy; pretending that the threat does not exist solves all problems about choice, ambitions and credibility once and for all. Having reached the conclusion that Swedish nuclear weapons were neither politically feasible nor within reach with limited defence spending, the Swedish government and the armed forces in the late 1960s simply abandoned the concept of nuclear attack in defence planning and lived happily ever after in the conviction that nuclear weapons and nuclear war was an international security concern, not a national defence problem.

The interaction of threats and protective strategies

Long-range weapons and high-value targets

On 23 March 1918 a sudden explosion rocked central Paris, soon followed by more detonations with a predictable regularity. However, no German Zeppelins or aircraft could be spotted, and the front, although closing in owing to the large German spring offensive, was still too far away for any known type of long-range artillery. Finally, a scout aircraft spotted from where the invisible attack on the capital had come: behind the front line the Germans had moved forward a huge and strange railway gun. The gun was a relatively small-calibre weapon, but the extremely long barrel gave the shell a range of 120 km, enough to target the whole of central Paris. Although the single 210 mm shells were unable to cause any widespread destruction, the loss of invulnerability in the French capital was annoying enough.

At these extreme ranges counter-battery fire was not feasible; the German gun could easily hit a large target like Paris but the lack of accuracy made a small target like the single German gun invulnerable to any French railway gun available. So, as long as the German offensive continued, and the French army was unable to push them back, the authorities in Paris had to resort to passive defensive measures like protecting the most valuable cultural objects with sandbags. The shelling of the French capital did not cease until the Allied armies had managed to halt the German offensive and push the Germans back over the Marne.

The Paris gun (or 'Big Bertha' as it was sometimes wrongly nicknamed) did not constitute a radically new weapons system for strategic warfare, but it highlighted the problems of protective strategies for high-value targets that were to play a crucial role throughout the twentieth century and beyond.

Developments in aerial warfare in the inter-war period changed and aggravated the threat against economic targets and population centres. The new technology seemed to favour the offensive employment of air power, an interpretation formulated in 1921 by the Italian general Giulio Douhet and widely accepted by proponents of aerial warfare.[2] Douhet argued that an aerial offensive alone could decide a war in swift initial blows. In 1928 he calculated that a war waged along these principles would be over in less than a month after 300 tons of bombs had been delivered over the larger cities and industrial centres of the adversary.[3] Against this formidable threat no defensive strategy was, according to Douhet, feasible. Anti-aircraft defence as well as fighter forces could never halt an air offensive. The only feasible defence strategy was thus a counter-offensive directed against the aerodromes of the opponent.

Most air forces in the inter-war period did develop postures and strategies influenced by the thinking of Douhet and like-minded proponent in other countries, such as the British Chief of the Air Staff Hugh Trenchard.[4] The assumed menace from air attack against cities and strategic nerve-centres was in the course of the 1930s highlighted by the German and Italian employment of bombers for terror strikes during the Spanish Civil War and the Italian use of aircraft for delivery of chemical weapons in the war in Abyssinia. At the outbreak of the Second World War, British authorities consequently calculated that an assumed German aerial offensive would result in casualties in the region of 100,000 dead and twice as many wounded in the first two weeks alone.[5] As it turned out, nothing like this materialized. Simple civil defence measures reduced casualties to a fraction of those expected, and not even in the devastated German and Japanese cities did the kind of collective collapse anticipated appear.

Miscalculations of the impact of city bombing were, however, only one explanation. Apart from civil defence, air defence proved far more effective than anticipated by the Douhet school. Radar enabled the British Royal Air Force to move from fighter point defence to a genuine air defence system for area defence. This system, and the favourable correlation of forces, did not mean that the RAF was able to stop German offensive air operations; the Battle of Britain was a battle of attrition rates. The same was the case with the vast layered air defence system established by the Luftwaffe to fend off the Allied bomber offensive from 1942 onwards. Although occasionally inflicting heavy losses on the bomber formations, the air defence was unable to effectively protect any single high-value target or substantially reduce the overall impact of the bomber offensive.[6]

The introduction of the German V-weapons highlighted the limitations of air defence measures. The aerodynamic V-1 made its first flight in December 1942 and its first guided flight in July 1943, when mass production of the weapon had already started. A total of 30,000 missiles is estimated to have been produced, of which approximately 8,000 were launched.[7] Owing to air reconnaissance and technical intelligence, the Allies knew of the coming V-1 offensive against London and could employ a number of countermeasures.[8]

The first strategy chosen by the Allies was pre-emption. In August 1943 British bombers attacked the German rocket research station at Peenemünde, but failed to hit the V-1 facilities. The next step was a massive air offensive directed against the launch sites constructed along the French Channel coast, the so called ski-sites. The ski-sites proved vulnerable to the concentrated air attacks (a total of more than 22,000 sorties between December 1943 and June 1944 knocked out 82 of 96 sites). The Germans, however, changed to smaller and thus more easily concealed launch sites of which approximately 150 were operational when the V-1 campaign was initiated in June 1944.[9]

On the whole, the attempts to pre-empt the V-1 campaign were ineffective; the campaign was delayed and started gradually, but more due to technical failures than the air attacks.[10] The British thus had to fall back on air defence and protective measures, trying to establish a layered air defence to prevent the missiles from reaching the assumed target, London. Initially a belt of anti-aircraft guns was placed halfway between the coastline and London, supplemented by a balloon belt closer to the target. Later, the gun belt was moved to the coastline to be more effective.[11]

The V-1 flew at 530–560 km/hr at an altitude between 600 and 900 metres, which made them vulnerable to intercepting fighter aircraft. The small size of the missile did, however, make target tracking difficult and the mix of fighter air defence and anti-aircraft artillery posed further tactical problems. Height and area separation was a major obstacle that could not be completely overcome, resulting in gaps in coverage or the risk of losses due to friendly fire.

Overall downing of observed V-1 missiles rose from less than 40 per cent to over 80 per cent. Of a total of around 7,500 V-1 missiles observed by the

British air defence, 52.3 per cent were downed. The air defence thus was far more effective than the pre-emptive strategy, but nevertheless failed to halt the V-1 campaign. On the whole, anti-aircraft fire (barrier defence) was more effective than fighters. In the defence of Antwerp only anti-aircraft artillery was employed and it destroyed 64 per cent of the incoming missiles.[12]

The V-2 posed a similar threat but created a different set of defensive problems for the Allies. Head of Scientific Intelligence R.V. Jones describes the advent of the V-2 attacks on 8 September 1944 in a slightly ironical light. The day before, the British authorities had announced the end of missile attacks, while in fact the Germans still controlled an area in southern Holland within 320 km of London (the V-2 missile's range). The impact of the first V-2 missile mocked the headlines that the 'Battle of London' was over.[13] A total of 1,190 V-2 (A-4) missiles were fired against the London area from September 1944 to April 1945. Of these, 1,115 impacts were registered, of which around 500 were in the London Civil Defence Region, killing 2,754 civilians (compared with 6,184 killed by five times as many V-1s).[14]

The ballistic V-2 missile was a far more advanced weapon than the V-1 'flying bomb' and compelled the Allies to search for countermeasures at both ends of the strategic spectrum. On the one hand, there was the strategy of pre-emption employed against the V-1 launch sites. But this was even more complicated against the mobile launchers of the V-2. The only practically feasible solution was to occupy the area from which the missiles were fired, something that was not achieved until the very end of the war.

On the other hand, air defence was not possible with the weapons systems available in 1944–5. Fighters were far too slow and artillery lacked range and accuracy. The only first-generation air defence missile system in advanced development was the German *Wasserfall*.[15] The only part of the air defence system with some relevance was the long-range radar that could spot the launch of the missiles and give the civil defence a few minutes' warning prior to a missile impact. As it turned out, it was not Allied countermeasures that averted the V-2 campaign but the limited destructive power of the warhead and the final German defeat.

Early Cold War strategies: monopoly, pre-emption and air defence

All this changed radically with the introduction of nuclear weapons. Not only did the balance between offensive and defensive air power shift, but also the entire basis for air defence. If one or two bombers could deliver a devastating attack on a city or an industrial complex, it was obviously pointless to make calculations in terms of attrition rate on an attacking bomber force.

Initially, nuclear weapons did not profoundly alter strategic premises. The United States enjoyed a complete nuclear monopoly and thus had no need for any other counter-strategy. The emerging main opponent, the Soviet Union, does not seem to have asserted the nuclear threat as a decisive factor until the post-Stalin era in the 1950s, and a major shift first appeared at the end of the 1950s.

The most radical form of defensive strategy was the line chosen by the US authorities during the Manhattan Project: the US would keep a monopoly on the new weapon, resist 'internationalization', and block attempts by other states to develop nuclear weapons. This motivated the extreme (but not very effective) secrecy, the unwillingness to give any information to the Soviet Union, and after the war also the exclusion of the British scientists from the project.[16] A specific aspect of this strategy was the measures taken by the US authorities in 1945 in ensuring that the leading German nuclear scientists were kept out of not only Soviet but also French hands.[17] In a similar fashion, extensive measures were taken to prevent producers of uranium ore from exporting this product to a third country.

The situation changed with the first Soviet nuclear explosion in 1949, shattering the US goal of ultimate non-proliferation. In the early 1950s the Soviet Union was assumed to have a small but rapidly growing nuclear stockpile, something that motivated a rapid increase in the US nuclear forces, first of all the Strategic Air Command. SAC originally planned for a bomber offensive along the same lines as during the Second World War directed against the Soviet industrial base. With the formation of NATO in 1949 this was supplemented with the task of halting a Soviet offensive into Western Europe. In 1950, following the Soviet nuclear test, SAC was assigned a third task, to blunt Soviet atomic capabilities. The blunting mission received highest priority from the Joint Chiefs of Staff.[18]

The 'blunting mission' was directed first of all against Soviet delivery systems and their bases but also nuclear research and production facilities. The target list immediately became long, since all major Soviet airfields had to be included.[19] The more the Soviet nuclear forces expanded, the more difficult the blunting mission became, until finally, in the 1960s, the point was approached (and passed) where a successful counter-force strike was no longer conceivable.

The blunting mission also highlighted the intricate question of prevention or pre-emption. For SAC the time available for preparations was of vital importance in the mid-1950s and a successful blunting mission required that the Soviet Union had not yet launched nuclear attacks on the Western powers. The more the blunting target was stressed, the more planning would thus enhance crisis instability.

The defensive strategies of the early Cold War focused on bomber aircraft as the primary means for delivery of nuclear weapons. Soviet bomber development was slow, owing to the lack of interest in strategic warfare under Stalin. Bomber development was speeded up in the mid-1950s with the introduction of all-jet and turboprop aircraft with intercontinental range, resulting in the famous 'bomber gap', a US intelligence misconception of the Soviet bomber production rate, fed by the earlier faulty estimate of the time needed for the Soviet Union to develop nuclear weapons.

Development of a Soviet air defence system was stepped up in the late 1940s with deployment of air defence radar along coastlines and in other forward areas, and the introduction of the MiG-15 jet fighter. However, a major reorganization appeared first in 1954 when a separate air defence command for the homeland, *PVO Strany*, was organized. From the early 1960s a second-generation air defence system was established, primarily to counter assumed US bomber paths over the polar area.

The 'bomber gap' served to fuel investments not only in US strategic air power but also in other parallel counter-strategies, first of all the establishment of extensive defensive systems for North America, also covering an assumed bomber route over the polar area. A chain of early-warning radar stations, the DEW (Distant Early Warning) line, stretching from Iceland to the Aleutian Islands, became operational in 1957. NORAD (North American Aerospace Defense Command) was established as a joint US–Canadian organization, using the strategic depth of the area. The DEW line was integrated in the command system, along with US and Canadian fighters and surface-to-air missiles for point defence, later supplemented by long-range Bomarc surface-to-air missiles. NORAD relied on an extensive use of nuclear warheads in the anti-aircraft mode.

The Soviet bomber threat anticipated in the mid- and late 1950s never materialized; the vast air defence system became obsolescent in the 1960s and 1970s and was gradually decommissioned and reoriented towards the role of early warning system also responsible for tracking ballistic missiles.

Sputnik and the missile gap: searching for the defensive options

Both the Soviet Union and the United States had captured technology and scientists from the German weapons programmes and initiated missile development based on the A-4 technology. In the US both the Air Force and the Army initially focused on early cruise missiles (Matador, Snark) and artillery rockets (Corporal, Honest John). The first ballistic missile to enter service was the Redstone in 1956, developed by the team headed by Wernher von Braun who had been employed by the US defence after the war. The Redstone was basically an improved version of the A-4, but still with approximately the same range. A successive system was the longer-range Jupiter, a development causing an intricate collision between the Army and the Air Force, the latter operating the similar Thor system.[20]

Soviet missile development also took its starting point in the A-4 missile, with the help of von Braun's less fortunate colleagues. The SS-1 Scud missile was, like the Redstone, a developed A-4, followed in the Soviet Union by the SS-2 Siebling. The rapid progress in the Soviet armed forces and missile ranges came with the military reforms under Khrushchev and the adaptation to war under nuclear conditions.

The sense of security for the North American continent created by the establishment of the extensive and layered defence system was shattered in one blow by the launching of the Soviet *Sputnik I* in October 1957. The 'Sputnik shock' had a profound impact on arms spending, strategy and R&D in the US. The hastily appointed scientific adviser, James Killian, described the effects in the following words in his memoirs:

> As it beeped in the sky, *Sputnik I* created a crisis of confidence that swept the country like a windblown forest fire. Overnight there developed a widespread fear that the country lay at the mercy of the Russian military machine and that our own government and its military arm had abruptly lost the power to defend the mainland itself, much less to maintain US prestige and leadership in the international arena.[21]

The link indicated by Killian between perceived homeland security and US prestige on the world arena is not without further significance as an ideological superstructure. The 'post-Sputnik' threat perception contained several components. First, there was the inflated anticipated threat from 'space bombardment', satellites carrying nuclear weapons to which the US would be open to attack. Second, the space launch demonstrated that the Soviets operated missiles with considerable lift capacity, missiles that also could be employed as ballistic missiles with an intercontinental range.

The 'missile gap' was the consequence of the Sputnik shock, but it was also deliberately fuelled by the Soviet leadership, who gave overstated accounts of the missile inventory and capacity of the Soviet long-range missiles, separated in 1960 into a new service, the Strategic Rocket Forces.

The advent of the intercontinental missiles created a defensive dilemma similar to the V-2 threat to Britain, although with the crucial difference that the multi-megaton warheads this time ruled out any passive protective measures for other than purely psychological purposes. There was simply a strategic and technological gap between an uncertain 'balance of terror' and strategies for pre-emption and the prospect of total destruction.

One immediate effect of the Sputnik and the fear of rapid Soviet developments in the field of ballistic missiles was the modification and development of the early warning and air defence system designed to counter a Soviet bomber threat. The NORAD air defence system contained the old Nike Ajax missile system that became operational in 1953, from 1958 onwards replaced by the Nike Hercules, with extended range and an alternative nuclear warhead. Following the Sputnik shock, a third version of the Nike system was rapidly designed: the nuclear-armed Nike Zeus, a three-stage missile with anti-ballistic capability. The Nike Zeus was designed as a dual ABM and anti-satellite (ASAT) system, the later called Program 505 and reflecting the fear of Soviet space-based systems.[22]

The Nike Zeus became the key missile system in the first US attempt to develop an ABM system and a parallel ASAT system. But in the ABM role the performance of the Nike Zeus system had a number of severe limitations. The system relied on mechanically operated radar that in practice could track only one target at the time. Another limitation was the slow speed of the missile, making it impossible to wait until a hostile warhead had entered the atmosphere and various decoys had been filtered away. Accuracy was not high; the question 'Can a bullet hit another bullet?' could only be answered positively if the effects of the one-megaton warhead were taken into account.[23]

Nike Zeus was thus not a credible answer to the Soviet missile threat, neither as this threat was anticipated in the late 1950s and early 1960s nor as it actually developed towards the mid-1960s. The ASAT system became more long-lived and illustrated the inherent tug-of-war between the US Army and Air Force. Nike Zeus was an Army programme and the revived interest in modifying the system for the ASAT role (mainly by extending the range) was viewed by critics as a way to save the ABM programme as well. Testing of the ASAT version started at Kwajalein atoll in 1963 and continued to 1966, when the system was finally decommissioned. Nike Zeus missiles were kept in operational readiness in the ASAT role until then. The Air Force opposed the further employment of Army ABM missiles in the ASAT role, since ASAT operations should in their opinion be a task for the Air Force. The US Air Force developed a parallel ASAT system, based on a modification of the Thor IRBM, known as Program 437, formally decommissioned in 1975.[24]

The great ABM debate: retreating up the ladder?

The major revision in Soviet thinking on nuclear war occurred around 1960 and resulted in a radically changed military planning for offensive operations in Europe under the conditions of a massive nuclear exchange.[25] The Soviet Union probably began early ABM developments in the mid-1950s, as a part of the emerging adjustment to nuclear war. US intelligence spotted the first signs of ABM sites around Moscow in late 1962, and in the November parade of 1964 the Galosh interceptor missile was for the first time displayed in public. In 1967 it was evident to US intelligence that the Soviets had embarked upon a massive build-up of ABM defences, not only around Moscow.[26] Like the Nike Zeus, the Galosh was a rather primitive first-generation ABM system with considerable shortcomings in the radar and tracking systems. It was basically a weapon capable of intercepting stray ballistic missiles under undisturbed conditions.[27]

The development of mutual ABM systems was hardly surprising, given the arms dynamics of the Cold War and the obvious strategic incentives at a stage in the nuclear arms race where neither side could count on even a theoretical possibility of disarming the opponent by a successful pre-emptive strike. And with the massive destructive power of the offensive weapons, no shelter or other civil defence programme was feasible to ensure survival.[28]

Realizing the limitations of the Nike Zeus system, the US Army in 1963 went ahead with the development of a follow-on system based on a more advanced missile, the Nike-X (Spartan). This system, under the name Sentinel, was intended for the defence of 25 selected US cities against limited ICBM attacks. The system, later renamed Safeguard and switched to defence of the US ICBM force, represented a second-generation ABM system. The phased-array radar systems were designed to be able to detect and track a vast number of objects simultaneously, and the shortcomings of the Nike Zeus missile were ameliorated by complementing the long-range Spartan missiles with short-range Sprint terminal interceptors. The components of the system were under development in the mid-1960s, and by 1967 a decision on production and deployment was approaching, further underlined by the rapid Soviet deployment of a first-generation system. The issue had intricate domestic political implications in the US, where an anticipated ABM 'missile gap' could be exploited in the 1968 presidential election.

The ABM development gradually became a public issue in the US in the mid-1960s with increasing criticism from scientists concerned about the arms race and sceptical of the feasibility of a national missile defence system.[29] In January 1967 the Johnson administration in the annual defence budget asked the Congress for funding for a component of an ABM system, while a decision to actually deploy such a system was postponed pending an attempt to negotiate with the Soviet Union.[30] Failure to acquire Soviet consent on the necessity to limit the construction of ABM systems at the Glassboro summit in June 1967 left the Johnson administration with the choice of expanding the strategic weapons programmes or starting an ABM programme that from a broader perspective of nuclear strategy did not make much sense. In a speech in September 1967, US Secretary of Defense Robert McNamara proposed what he called a 'light' ABM system. McNamara argued that an ABM defence against the Soviet nuclear inventory would be destabilizing, and he focused instead on the assumed limited threat from future Chinese ICBMs in the 1970s. The ABM debate thus got a somewhat contradictory start, with proponents of the system focusing on the main arms control arguments against the system.[31]

The focus on the Chinese threat, with the protection against Soviet attacks more as a by-product, was motivated by the domestic political setting rather than strategic considerations. This was even more apparent after the Nixon administration had entered office in 1969; the Sentinel system was now renamed Safeguard and although consisting of the same components, was switched from city defence to protection of the US land-based retaliatory forces against a direct attack by the Soviet Union, with defence against a Chinese attack as a side effect. The switch from city defence was explained with reference to the unfeasibility of that task: 'In view of the magnitude of the current Soviet missile threat to the United States, and the prospects of future growth in quantity and quality, we have concluded that a defence of our populations against that threat is not now feasible.'[32]

The critical debate among the experts had during 1967 spread out into the public domain, fuelled by the growing critique against the Vietnam war, the military system and the establishment in general. To this was added a grass-roots protest rapidly spreading over the US against the construction of ABM sites for missiles armed with nuclear warheads to be detonated in the atmosphere.[33] Unique for the ABM debate was the split in the US Senate, where a large minority opposed funding for the ABM system.

At the level of nuclear strategy and arms control the controversy focused on the feasibility of the proposed system and its assumed effects for survivability, crisis stability and arms control. The survivability in general terms was not an issue, once the system had been designed to protect the ICBM bases; instead the main controversy developed around the assumed effect in a crisis and at various hypothetical stages of a nuclear exchange. Broadly, the proponent's main arguments on this point were:[34]

- Missile defence would enhance deterrence by introducing a further uncertainty for the attacker and reduce incentives for a first strike.
- The ABM system was necessary to offset the assumed Soviet superiority in offensive weapons.
- Defensive systems were morally preferable to offensive.

Against this the opponents argued:

- No Soviet threat motivated the ABM system.
- The ABM system proposed would not have the assumed effect owing to technical limitations, political constraints affecting a decision to launch missiles, and sensitivity to various counter-strategies.
- The ABM deployment would increase the arms race.
- The US and Europe would be strategically decoupled.[35]

The two sides dug in on their respective positions, while the ABM system increasingly moved over from the domain of strategic choices to the negotiation table as one of several pawns in a larger disarmament game.

From arms race to treaty: creating the deterrence universe

The concept of deterrence as initiated or rather formulated by McNamara in 1967 constituted a major change in nuclear thinking. Deterrence, according to McNamara, was not the surrender of defensive strategies but rather an expansion, adding a political aspect to pre-emption: 'It is our ability to destroy an attacker as a viable twentieth century nation that provides the deterrent, not our ability to partially limit damage to ourselves.'[36] Formulating a second strike capability as the ultimate defence became the cornerstone of détente and nuclear thinking until the early 1980s.

The Sentinel programme was, however, not only motivated in terms of nuclear strategy but also in the context of arms control. In the 1960s the decisions on nuclear weapons systems started to shift from a strategic rationale to an arms negotiation rationale. In 1970 the US Secretary of Defense Melvin Laird thus motivated funding of the ABM system in the latter terms: 'An orderly, measured, flexible and ongoing Safeguard defence program will help maintain our relative positive position in SALT and improve the chances for a successful outcome.'[37]

At the Glassboro summit in June 1967, President Johnson had in vain tried to convince Prime Minister Kosygin that it was a wise thing to enter negotiations to limit the development of ABM systems as a way to prevent an arms race. McNamara seconded Johnson, but an infuriated Kosygin maintained that defence was moral while offence was immoral.[38]

Between 1967 and 1969, when negotiations on ABM limitations were initiated as a part of the broader Strategic Arms Limitation Talks, the Soviet attitude underwent a profound change. From complete rejection of any idea of limiting ABM deployment, the Soviet leadership now accepted the idea of limitations, possibly influenced by the US Sentinel/Safeguard programme and the considerable limitations of the first-generation Soviet ABM defence system. Doubts about the possibility of effectively blocking a determined nuclear attack seem to have influenced the Soviet leadership in the late 1960s, much in the same way as McNamara's ideas in the US.[39]

When the talks on strategic weapons opened in 1969, the Soviet delegates were the most eager to discuss the ABM issue. In April 1970 the US delegation presented a suggestion to limit each side's ABM capacity to one system to protect the capital (or National Command Authority). It is unclear whether or not this was a tactical move, but to the surprise of the US the Soviet side accepted the proposal and firmly advocated the NCA-only concept. With the US Congress approving the ICBM protection deployment, this put the US side in an unfavourable position, since it was highly unlikely, given the ABM debate, that a city-protection system would ever be approved.[40]

In the end an arms control package, including what was to become known as the SALT-I accord and the ABM treaty, was signed in May 1972. In the initial form the treaty ended up as a compromise, allowing for both parties to deploy an ABM system at two sites each. In the 1974 protocol added to the treaty, both sides agreed to formalize their existing deployment, the US at Grand Forks Air Force base and the Soviet Union around Moscow.

A strategic deadlock had emerged, similar to the one experienced between the major battle fleets in the First World War, but now of a seemingly more permanent nature. Deterrence offered a solution to the unsolvable strategic dilemmas of the Missile Age, but a solution influenced both by the available and foreseeable technologies and the specific conditions in world politics in the late 1960s and 1970s.

The origins of Star Wars: strategic drift from deterrence

With the ABM treaty the issue of ballistic missile defence systems seemed to be solved permanently.[41] However, the firm link established in the parallel SALT and ABM treaties between offensive and defensive systems meant that the issue of strategic defensive systems and concepts was bound to surface as soon as the specific circumstances for the SALT process were no longer at hand.

The origins of the return of the nuclear defence issue can be traced to the late 1970s, when the official US policy on arms control negotiations was increasingly criticized for underestimating a growing threat from Soviet strategic forces. The emerging controversy was focused on conventional and strategic weapons, and so the ABM system continued to be a non-issue. The Reagan administration, coming into office after the 1980 election, pursued a policy of broad rearmament to face what was described as a growing Soviet threat under disguise of détente and arms control. The single most decisive step was the non-ratification of the SALT-II agreement and the withdrawal from the SALT process (the NATO double-track decision on European strategic nuclear weapons had been taken already by the Carter administration).[42]

The issue of defence against nuclear weapons was at first not addressed by the Reagan administration even though one clear ambition was to make nuclear weapons 'thinkable' again. The SIPRI director, Frank Blackaby, describes the matter in the following way:

> This revival of the idea of ballistic missile defence (BMD) was not the result of a careful reappraisal of strategic doctrines. It did not emerge from a process of inter-agency consultation within the US bureaucracy. The idea of BMD had for a long time been out of the mainstream of US strategic thinking. There were of course research programmes in being; but there was no forceful intention of working towards the development of a BMD system.[43]

The ideas that in 1983 were transformed into the Strategic Defence Initiative originated or at least were channelled through the retired general Daniel O. Graham, who served as adviser to Ronald Reagan in the 1976 and 1980 elections. After the 1980 election, Graham founded the private lobby group High Frontier. Graham wanted a profound change in US security: 'I was among those insisting that the only viable approach for a new administration to cope with growing military imbalances was to implement a basic change in US grand strategy and make a technological end-run on the Soviets.'[44] The defence initiative proposed by High Frontier had distinct messianic dimensions: 'The United States is faced with a historic, but fleeting, opportunity to take its destiny into its own hands. The ominous military and economic trends which today beset the peoples of the Free World can be reversed, and confidence in the future of free political and economic systems can be restored.'[45] More concretely, High Frontier advocated an exploration of space technologies in general and space defence in particular. A multilayered space-based defence system should 'nullify the present and growing threat to the US and its allies' and 'replace the dangerous doctrine of Mutual Assured Destruction (MAD) with a strategy of Assured Survival'.[46]

The veteran scientist Edward Teller and Admiral James Watkins persuaded the Joint Chiefs of Staff to raise the issue of defensive technologies with the President in February 1983. The question of space-based systems was not presented in a concrete form, but President Reagan was enthusiastic and a small group of advisers were assigned to add a five-minute section to a speech the President was to deliver in March 1983, later known as the 'Star Wars speech'. According to the interpretation of Blackaby, the release of the idea was deliberately done in secrecy to avoid the triple threat of Congress, the press and the bureaucracy. The Secretary of State was given two days' warning, the chief Pentagon scientist only nine hours.[47]

In his speech to the nation, President Reagan proposed a strategy that would aim at a strategic shield that would render nuclear weapons impotent and obsolete. The Reagan vision, although bypassing other means of nuclear delivery than ballistic missiles, promised a path that would turn the clock back to the situation prior to the advent of the Soviet missile threat when the US for all practical reasons enjoyed immunity to direct attacks. SDI would with this interpretation re-establish the US as a true superpower, second to none.

The critics of the SDI programme, initiated in late 1983, partly voiced concerns similar to the ABM debate in the 1960s. First there was the question of technological feasibility. SDI was a concept of possible future weapons technologies, not like the Sentinel/Safeguard an existing system, which had proceeded past the development and test phases. The whole idea of a space-based defence with certain characteristics and performance was hypothetical. As in the ABM debate, critics focused both on the possible negative effects on crisis stability and the threat from a renewed strategic arms race, one that would now extend into space. The debate among specialists did, however, contain a new and in many ways dominating aspect, the interpretation of the ABM treaty and the effects of unilateral US diversions from the treaty.[48] Possible Soviet violations, first of all the large ABM radar constructed at Krasnoyarsk, thus became a key issue, which it most likely would not have become under other circumstances.

Although criticized, the SDI did not develop into a major political let alone public controversy, as was the case with ABM. The SDI debate was about complex technological issues and research policy, not the stationing of nuclear weapons in the neighbourhood or the prospect of nuclear detonations overhead. Not for nothing had General Graham focused on the non-nuclear aspect of the defence against the hostile nukes.

From SDI to BMD: weapons in search of a strategy?

The SDI programme did not, as its opponents had warned, stumble into a morass of rising costs, disastrous arms control effects and unforeseen and unsolvable technological problems. Instead the programme unavoidably became one of the victims of the fading Cold War and the subsequent collapse of the Soviet state. The link between offensive and defensive weapons, established with the ABM treaty, now worked the other way; with prospects of rapid dismantling of large proportions of the ex-Soviet strategic nuclear inventory, basically all the original arguments for the SDI programme had become irrelevant. There was no need for a costly and technologically unproven initiative to save the United States from the trap of mutual assured destruction when the political rationale for mutually threatening force postures between Russia and the US no longer existed. If the 'Reagan revolution' was successful in the overriding aim of breaking the back of the Soviet arms economy, then the original strategic reasons for SDI were also gone.

In the late 1980s the Bush administration continued to ensure full support of the SDI programme, based on the strategic rationale of the system and the US interpretation that the programme was consistent with the ABM treaty. In April 1989, Secretary of Defense Dick Cheney in a testimony before the House Armed Services Committee declared, 'the goal of the Strategic Defence Initiative remains unchanged'. However, a 'restructured program' would continue towards development of the most promising technologies, as 'Brilliant Pebbles'.[49] In February 1990, President Bush focused on more limited objects for strategic defences: 'beyond their contribution to deterrence, they underline effective arms control by diminishing the advantage of cheating. They can also defend us against accidental launches – or attacks from many other countries that, regrettably, are acquiring ballistic missile capabilities.'[50] This marked the reintroduction of the threat from the 'new' opponents with WMD and missile capability that were to dominate the developments from 1991 onwards.

The 1991 Gulf War underlined and accelerated this shift in strategic focus for missile defence from the Cold War nuclear balance to the new regional threats to US interests in the post-Cold War world. At the same time, Iraqi Scud (Al Hussain) attacks against Israel and Saudi Arabia and intercepts by Israeli and US Patriot missiles supplied the first test ever under combat conditions of a ballistic missile defence system, illustrating the difficulty of theoretical assessments of actual performance.[51] As a response to the Gulf War, and the search for a mission for the key components of the SDI programme, the Bush administration redefined the programme as GPALS (Global Protection against Limited Strikes), a name that highlighted the shift from comprehensive to limited threats. However, the scaled-down ambitions did not decisively alter the technological content of the former SDI programme. GPALS was followed by a US diplomatic offensive towards Russia to convince the new Russian leadership of the necessity of altering the limitations of the ABM treaty.[52]

The Clinton administration initially took a different stand on missile defence, downgrading it and in reality dissolving the remnants of the original SDI programme. GPALS was replaced by a less ambitious focus on Theatre Missile Defence (TMD). This was motivated partly with reference to the hypothetical ballistic missile threat (at least to the North American continent), partly as a consequence of supporting the ABM treaty with the aim of compelling the Russians to proceed with reduction in the offensive nuclear forces.

In 1995 the National Intelligence Estimate (NIE) came to the conclusion that 'no country, other than the declared nuclear powers, will develop or otherwise acquire a ballistic missile in the next 15 years that could threaten the contiguous 48 states and Canada'.[53] This conclusion, echoing earlier periods of assumed strategic immunity, became the target of a growing opposition, accusing the NIE of being a politicized intelligence estimate, and in 1998 a group headed by the former Republican Secretary of Defense challenged the assessment as incorrect and writing down the potential threat.[54] Six weeks after the report was presented, North Korea launched a three-stage rocket in an attempt to place a satellite into orbit.

Pressed by domestic critics and a growing support for a missile defence system, the Clinton administration in January 1999 decided to allocate funding for deployment of a limited National Missile Defence (NMD) and in June the President signed the Republican-inspired NMD Act. The purpose of the defence system was now defined as protection against limited ballistic missile attacks, whether accidental, unauthorized or deliberate. A new NIE in September 1999 generally confirmed the findings of the Rumsfeld Commission.[55]

The decision to deploy a defence system opened the missile defence issue for the third time. This time, critics referred not only to the Russian reaction and the hazards of abandoning the ABM treaty but also to the reaction of the Chinese; a national missile defence might threaten to make the ageing Chinese ICBM force inadequate, provoking China into a hostile political reaction or increased development of its strategic nuclear forces. China thus appeared as a borderline case, a power that the missile defence (officially) was not designed against – as compared to the ABM systems in the late 1960s – but that anyway might be affected by the relatively limited and unsophisticated arsenal of delivery systems.[56]

The nearly two-decades-old issue of the interpretation of the ABM treaty was finally pushed aside by the George W. Bush administration in December 2001, when the Russians were formally notified of the US intention to withdraw from the treaty six months later in June 2002. At that time, the build-up had started for the first preventive war explicitly legitimized with the need to deprive a state once and for all of the capability to use or develop WMD and their delivery systems.

Conclusions: 11 September and the leap to pre-emption

Looking back on the nuclear age following the Second World War, the span of possible counter-strategies for coping with weapons of mass destruction and their means of delivery was limited and the choices in many cases predictable, with technology and political feasibility as the fundamental variables. The Cold War in this respect represents a kind of kaleidoscopic experience, where all solutions in numerous combinations were tried, sometimes over and over again, as was the case with systems for ballistic missile defence. And as it turned out, nuclear deterrence and successive balanced arms limitations were not a final outcome, only a political choice under certain temporary circumstances.

After the end of the Cold War the pattern has changed, more dramatically than most had expected. But the strategic choices still remain within the general framework of a span from reactive to proactive measures. Deterrence through balance has lost its predominant role, compared to the high tide in superpower bilateralism from the mid-1960s up to the decline of the ex-Soviet military apparatus. In a unipolar world, the remaining hegemonic superpower has no need for deterrence through balance, on the contrary. From the historical perspective the resumed emphasis on Ballistic Missile Defence thus represents both a logical consequence of an emerging unipolar security system and a return to strategic concepts – and strategic problems – from the early phases of the Cold War. But, as the terrorist attacks of 11 September 2001 illustrated, the interaction between strategic offence and defence is an ongoing and occasionally revolutionary process with no technological final solution within reach. Missile defence commands the middle ground in the strategies for coping with threats from WMD, but only as long as the mid-twentieth-century invention, the ballistic missile, is the means of delivery.

The terrorist attacks of 11 September 2001 against the continental US had a profound impact on US threat perceptions, self-images and security policy. In reflecting on the nature of the threat, it might be illustrative to recall the strategic concept or rather vision formulated by Douhet in the 1920s: a blow against a nation's most vulnerable and vital centres would be more devastating than an attack on the armed forces, since these would be demoralized and unable to act in the defence of the motherland. Of course, this did not come about, neither in the armed conflicts of the 1930s and 1940s nor in the massive military response launched by the United States. Still, there is some irony in the fact that a handful of semi-skilled terrorists in a few hours managed to create the kind of collective psychological shock that no bomber offensive so far had resulted in.

Notes

1 See for instance Ashton B. Carter, 'How to Counter WMD', *Foreign Affairs*, Sept./Oct. 2004. Carter, who served as Assistant Secretary of Defense in the Clinton administration, argues that a more focused US counter-proliferation policy is the key strategy to counter the transformed WMD threat.
2 Giulio Douhet, *Il Dominio dell'Aria*, Rome, 1921.
3 P. Vautier, *Die Kriegslehre des Generals Douhet*, Berlin, 1935. Douhet had published his 300 ton estimate in the Italian *Rivista Aeronautica*, July 1928.
4 See Neville Jones, *The Beginnings of Strategic Air Power: A History of the British Bomber Force 1923–39*, London: Frank Cass, 1987. Douhet's original book was translated into German and published in 1935.
5 Freeman Dyson, *Weapons and Hope*, New York: Harper & Row, 1983, pp. 17 f.
6 The actual impact of the Allied bomber offensive was more limited than anticipated during the war.
7 Rudolf Lusar, *German Secret Weapons of the Second World War*, London: Spearman, 1959; Kenneth P. Werrell, *The Evolution of the Cruise Missile*, Waxwell Air Force Base: Air University Press, 1985.
8 A detailed account of the intelligence on the V-1 is given in R. V. Jones, *Most Secret War: British Scientific Intelligence 1939–1945*, London: Coronet, 1979.
9 Werrell, op. cit., p. 44.
10 The Germans had planned to start the V-1 campaign with a massive attack on 11 June 1944, but of the first salvo only nine left the launch sites and none reached Britain. Of a second salvo of ten only four reached Britain. The first impact of a V-1 in London was at 4:18 a.m. on 13 June. Ibid.
11 Ibid.
12 Werrell, op. cit., p. 54.
13 Jones, op. cit., p. 578.
14 Ibid.; Werrell, op. cit., p. 60.
15 Lusar, op. cit.; Leslie E. Simon, *German Research in World War II*, New York: John Wiley, 1947. *Wasserfall* was designed as an infra-red homing anti-aircraft missile and did not have ABM capability.

16 Margaret Gowing, *Britain and Atomic Energy 1939-45*, London: Macmillan, 1965, and *Independence and Deterrence: Britain and Atomic Energy 1945–1952*, vol. 1, London: Macmillan, 1974.
17 On the ALSOS mission, see Samuel A. Goudsmit, *Alsos*, Woodbury, NY: AIP Press, 1996, and Jeremy Bernstein, *Hitler's Uranium Club: The Secret Recording at Farm Hall*, Woodbury, NY: American Institute of Physics, 1996.
18 David Alan Rosenberg, 'A Smoking Radiating Ruin at the End of Two Hours: Documents on American Plans for Nuclear War with the Soviet Union 1954–55', *International Security*, 6(3), Winter 1981–2.
19 Ibid.
20 See also David Miller, *The Cold War – a Military History*, New York: St Martins Press, 1998, p. 96.
21 James R. Killian, *Sputnik, Scientists, and Eisenhower: A Memoir of the First Special Assistant to the President for Science and Technology*, Cambridge: MIT Press, 1982, p. 7.
22 For the ASAT program see Paul B. Stares, *The Militarization of Space: US Policy, 1945–85*, Ithaca, NY: Cornell University Press, 1985, pp. 117–120, and Thomas Karas, *The New High Ground: Strategies and Weapons of the Space Age War*, New York: Simon & Schuster, 1983.
23 D.G. Brennan and Johan J. Holst, 'Ballistic Missile Defence: Two Views', *Adelphi Papers*, no. 43, November 1967.
24 Stares, op. cit., pp. 120–128.
25 The focus on a future war as a large-scale nuclear confrontation was clearly stated in the book *Military Strategy*, by Marshal V.D. Sokolovsky, published by the Soviet Ministry of Defence in 1962 and published in English the year after (New York: Frederic A. Praeger, 1963). Declassified Warsaw Pact planning documents from the early 1960s confirm that this strategic view imbued operational planning at all levels.
26 Kerry M. Kartchner, 'Origins of the ABM Treaty', in James J. Wirtz and Jeffrey A. Larsen (eds), *Rockets' Red Glare: Missile Defenses and the Future of World Politics*, Boulder, Colo.: Westview Press, 2001; also Lawrence Freedman, *US Intelligence and the Soviet Strategic Threat*, Princeton, NJ: Princeton University Press, 1986.
27 According to US official sources the Galosh system was operational until the mid-1980s when it was upgraded with the new missiles SH-04 and SH-08. See *Soviet Military Power 1987*, Washington: US Government Printing Office, 1987.
28 This transition from early nuclear war fighting into the increasingly bizarre world of mutual assured destruction is well described in the study of British war planning by Peter Hennessy, *The Secret State: Whitehall and the Cold War*, London: Allen Lane, 2002.
29 The start of the debate among academics was an article by Herbert York and Jerome Wiesner, 'National Security and the Test Ban,' in *Scientific American*, October 1964. See Kartcher, op. cit., p. 34.
30 John W. Finney, 'Paper 1. A Historical Perspective', in Walther Stützle, Bhupendra Jasani and Regina Cowed (eds), *The ABM-Treaty: To Defend or Not to Defend?*, SIPRI/Oxford University Press, Oxford, 1987.
31 Ibid., pp. 31–34.
32 Kartchner, op. cit., p. 25, quoting statement by Secretary of Defense Melvin Laird to the Senate Armed Service Committee, 20 February 1970.
33 Proponents of the ABM system sourly noted that the US Army had for years deployed nuclear-tipped Nike Hercules missile batteries around major cities

without anyone protesting, since the Army had not spoken openly about the deployment. See Finney, op. cit., p. 34.
34 Kartchner, op. cit., pp. 35–36; Brennan and Holst, op. cit.
35 See argument of Holst, ibid.
36 Kartchner, op. cit., p. 32, quoting statement by McNamara to the Senate Armed Service Committee, 25 January 1967, reprinted in *Documents on Disarmament 1967*.
37 Kartchner, op. cit., p. 25.
38 Robert S. McNamara, *Blundering into Disaster: Surviving the First Century of the Nuclear Age*, New York: Pantheon, 1986, p. 57.
39 Kartchner, op. cit., p. 41.
40 Ibid., pp. 26–27.
41 One indication is the almost complete lack of interest in the details of the treaty when it was ratified by the US Senate, although the Senate had for years been the centre of gravity in the domestic ABM debate.
42 The Reagan administration, while initiating a massive nuclear rearmament programme, proposed talks on strategic arms reduction (START) instead of arms limitation (SALT). Owing to deteriorating US–Soviet relations in the early 1980s, negotiations first started in the mid-1980s.
43 Frank Blackaby, 'Space Weapons and Security', *SIPRI Yearbook 1986*, Oxford: SIPRI/Oxford University Press, 1986, p. 82.
44 Daniel O. Graham, *The Non-Nuclear Defense of Cities: The High Frontier Space-Based Defense Against ICBM Attack*, Cambridge, Mass.: Abt Books, 1983, p. vi.
45 Ibid., p. 1.
46 Ibid. It is worth observing that Mutual Assured Destruction would be replaced by Assured Survival, not *Mutual* Assured Survival.
47 Blackaby, op. cit., p. 83.
48 For the controversy over legal interpretations see Gerard C. Smith, 'The Treaty's Basic Provisions: View of the US Negotiator', and Vladimir Semenov, 'The Treaty's Basic Provisions: View of the Soviet Negotiator', both in Stützle *et al.*, op. cit.
49 *Strategic Defence Initiative: SDI Chronology 1983–1990*, United States Information Agency, 19 March 1990.
50 Ibid. Speech by President Bush at the Lawrence Livermore National Laboratory, 7 February 1990.
51 For the debate over the Patriot performance see Theodore Postol, 'Lessons of the Gulf War Experience with Patriot', *International Security*, Winter 1991–2. It is today generally accepted that the Patriot system was ineffective against incoming ballistic missiles.
52 See Robert Joseph, 'The Changing Political-Military Environment', in Wirtz and Larsen, op. cit.
53 Wyn Q. Bowen, 'Missile Defence and the Transatlantic Security Relationship', *International Affairs*, 77(3), 2001.
54 Ibid.
55 Ibid.
56 See Tim Young and Claire Taylor, 'Ballistic Missile Defence', Research Paper 03/28, House of Commons Library, 2003.

2

US MISSILE DEFENCE

Technological primacy in action

Tarja Cronberg

Introduction

The decision by President George W. Bush to deploy a national missile defence system is historic. For the first time a system covering a whole nation will be constructed. Missile defence[1] is expected to protect US citizens, in all the 50 states, from an attack by a few tens of missiles from not only North Korea and the Middle East but also from accidental launches from Russia or China. It would be immoral, many argue, not to construct a defence system when it is becoming technologically feasible. 'We will defend ourselves with our technology' was a point made by George W. Bush in his inaugural speech.

Missile defence is thus not only about defence but also about technology. To see missile defence only as the technological answer to the rogue state missile threat gives a very limited picture of the stakes involved. If the goal was only to protect the US homeland against these threats, there would be other more effective and above all cheaper ways to reach this goal, such as negotiations, enforcing the Missile Technology Control Regime and other international conventions, sanctions, or even pre-emptive strikes. Missile defence and the development of technologies to make it possible form a long-term trajectory within the US military-industrial establishment. In fact, one could maintain that in the history of missile defence, threats have been constructed to match the politics of technology.

There are greater threats, I argue, to the US than a ballistic missile launched from a rogue state or accidentally from Russia or China. These threats are related to space and the vulnerability of space systems. There is, in the US, a fear of a Pearl Harbor in space. Consequently, space has become a top US security concern. The technologies needed for the protection of space assets are largely the same as those needed for missile defence. US technological policy and the very special role that military R&D plays within this context form another part of my explanation for why the US is willing to challenge the international arms control regime and in the long run even international stability in pursuing its national interests.

In this chapter, theoretically balancing between missile defence as a technological trajectory and as a social construct, I show how the US security interest in space, the needs of the military R&D system and the ballistic missile threat from rogue states all converge in the development of space-based sensors and interceptors, critical components of missile defence. My argument is thus: to understand the American missile defence plans, the issues of space and military R&D have to be included. Accordingly I start out by outlining the overall continuities of the missile defence issue. This is then contrasted with the variation in threats that the system has been supposed to counter. To explain that apparent inconsistency between continuation and variation and especially the current developments, I turn to the dual elements of US space and industrial policy.

A technological trajectory

Although there has been little public debate about missile defence since Ronald Reagan promoted the Star Wars concept, US spending on missile defence since 1986 has been surprisingly stable. Yearly spending has been in the range of $4–5 billion a year (constant 2001 dollars), although in 1994–5 it reached an all-time low of $3 billion (during both years). While spending has been stable, technological priorities have oscillated. In the 1980s a comprehensive national missile defence (NMD) and supporting technologies were the highest priority. In the 1990s the main focus was on theatre missile defence.[2] The distinction between a national and a theatre missile defence is one of scale. Theatre missile defence (TMD) is aimed at protecting US troops and allies against attacks by short-range missiles such as the Scud during the Gulf War. These systems do not threaten the deterrent of other nations. National missile defence systems, on the other hand, protect both cities and missile launch sites from long-range ballistic missiles. In doing so they undermine the deterrent of the enemy.

While research programmes were initiated earlier, the first major step to national missile defence was taken when the Johnson administration proposed building the Sentinel system in the late 1960s. Nuclear-tipped interceptor missiles were to be placed at 15 sites around the country including ten near major metropolitan areas. As people living near the planned sites rebelled, Sentinel was abolished and replaced by a Safeguard system. Here the same technology would be used to protect intercontinental ballistic missile (ICBM) sites.

The Anti-Ballistic Missile (ABM) treaty in 1972 was a milestone. It banned all forms of national missile defences in the US and Soviet Union against long-range missile attacks. However, it permitted both countries to operate two missile defence systems, one at its national capital and the other at an ICBM site. In 1974 the number was cut to one. Russia maintained the site around Moscow and the US defended an ICBM site. Reagan's 1983 Star Wars speech, however, again destabilized the situation. The US was to launch a new Strategic Defence Initiative (SDI), which also included defence of the US with space-based systems. The SDI, a strong platform of research, was never actually deployed as a defence system despite its high political profile. During the following Bush and Clinton administrations the focus turned to more limited objectives reflected in the Global Protection Against Limited Strikes (GPALS) programme. Despite these limited objectives, Bill Clinton was the president who signed the National Missile Defense Act. The actual decision to deploy a defence system was, however, subject to additional tests and deferred to the Bush administration.

In his 2000 presidential campaign Bush expressed interest in robust defence systems that included space-based weapons.[3] In December 2002 President Bush made the decision to deploy an initial midcourse missile defence capability consisting of 20 ground-based interceptors employing exo-atmospheric kill vehicles, three Aegis-class cruisers and an unspecified number of Patriot interceptors supported by various early warning and command and communications capabilities. In 2004 six ground-based interceptors would be placed at Fort Greely, Alaska, and another four at Vandenberg Air Force Base, California. By the end of September 2004 at least some of the interceptors at Fort Greely would be placed on alert. Another ten ground-based interceptors would be added at Fort Greely in 2005. This would mean that the US could defend itself against limited attacks from North Korea already in 2004 and against attacks from the Middle East by 2005, given that the radars at Thule and Fylingdales are upgraded. President Bush could thus deliver on his 2000 campaign pledge to deploy a national missile defence system.

Missile defence systems offer a set of technological choices. When an enemy fires an ICBM carrying a WMD warhead, a counter-missile is fired to destroy the incoming missile by hitting it or targeting it with energy weapons. The first alternative is the boost phase, i.e. shortly after launch. In theory this would be the easiest, as the missile would be destroyed before it deploys warheads or decoys. Furthermore, the ICBM is moving rather slowly and the rocket plume is easy to locate. On the other hand, early detection is difficult, and the time limit to reach the target is only up to five minutes after launch.

The Clinton administration proposed to build a midcourse interceptor using the so-called 'hit to kill' technology. The target is on its midcourse when the defence missile would launch its 'kill vehicle' to destroy the incoming warheads by hitting them. Time is less critical and, depending on the launch site, the flight time may be up to 20–30 minutes. Problems arise, as the enemy missile is likely to consist of warheads as well as decoys, which outside the atmosphere travel at the same speed. The warheads have to be destroyed (and the decoys distinguished from the warheads) before they re-enter the atmosphere. To target the incoming warhead in the final phase, when it re-enters the atmosphere, is also an option. In this case there is a risk of potential fallout depending on whether the warhead carries nuclear, chemical or biological weapons.

Launch facilities may be land, sea, air or space-based. The US Navy has developed a theatre-wide missile defence system, which could be expanded to become a national midcourse defence system as well. The Pentagon also has a programme for an airborne boost-phase defence. Aircraft would carry airborne lasers generating an intense beam of light, which would weaken the incoming rocket's outer surface causing it to rupture.[4] It is projected that these different systems will be combined into a more layered 'thick' missile defence. The midcourse defence now planned could be supplemented by a boost-phase defence based near the potential enemy country including both ship and airborne missile defences. In the future, space-based interceptor missiles could track an attacking ballistic missile in the boost phase, combined with ground-based interceptors in the midcourse as well as the final phase.

A national missile defence system, however, gives the attacker an advantage. The defence must commit itself to a specific technology and architecture before the attacker does. This allows the attacker to tailor countermeasures to a specific defence system. For hit-to-kill interceptors, which destroy the incoming missile by colliding with it, there is little margin for error. The development of countermeasures is not just a theoretical possibility, but rather something the countries possessing intercontinental ballistic missiles have been engaged in for a long time. In fact, it can be assumed that any country possessing the ability to launch this kind of missile will also have the technological capabilities of developing countermeasures. Furthermore, the defender has a much more difficult time in detecting the development of countermeasures.

The Union of Concerned Scientists and the Security Studies Program at the Massachusetts Institute of Technology carried out, in 2000, an in-depth study of countermeasures.[5] The study concludes that the national missile defence system under deployment could be easily defeated by simple countermeasures:

> Many highly effective countermeasures require a lower level of technology than that required to build a long-range ballistic missile (or nuclear weapon). The United States must anticipate that any potentially hostile country developing or acquiring ballistic missiles

> would have a parallel program to develop or acquire countermeasures to make those missiles effective in the face of US missile defenses. Countermeasure programs could be concealed from US intelligence much more easily than missile programs, and the US should not assume that a lack of intelligence evidence is evidence that countermeasure programs do not exist.[6]

The report recommends that 'a deployment decision should be postponed until the system has been tested successfully against realistic countermeasures'. On the other hand, the authors recognize that this will not be the case and that such tests are not even planned.

This short historical view of missile defence shows that the general idea of defending against ICBM's has – in different variations – for long been a subject in American security policy, even though the technological viability of the idea has been questioned.

Constructing threats

The development of missile defence systems is accordingly a long-term technological trajectory with fairly stable financing. In contrast to this stable trajectory, the question 'Against what should the missile defence work?' has had many different answers. In fact, the missile defence as an arena for research and technical development has been amazingly adaptable to any images of threat that have been put forth by the US military-political system. In terms of the sociology of technology the missile defence thus becomes a hybrid between a technological trajectory, a concept used in innovation economics,[7] and a social construction of technology,[8] which claims that technology is malleable and that there are always choices to be made. In order to understand this interplay between a more deterministic trajectory and the social construction of arguments, I shall shortly review the threat images through the history of missile defence.

In the beginning the main goal of missile defence was defence against Russian missiles. As the Soviet Union started to build an ABM shield over Moscow in 1964 the early US missile defence systems were geared to limit the Soviet second-strike capacity. However, as the arms race dynamics of the Cold War entailed a steady growth in both number and sophistication of the Soviet ICBMs, it became clear that a missile defence system based on present technology would in any case be saturated by sheer numbers. Thus, as the concept of deterrence through Mutually Assured Destruction became institutionalized as the guarantee of stability between the US and the USSR, missile defence was seen as a potentially destabilizing element. Accordingly, the ABM treaty set the limits for missile defence development given that both countries had nuclear arsenals large enough to penetrate each other's missile defences. This changed of course as Reagan introduced the SDI, the goal of which exactly was to question the stability of deterrence by means of reintroducing missile defence. Today Russia's more than 5,000 warheads provide a deterrent both with or without the current missile defence plans, although some sources claim that at some point, given a large number of interceptor missiles, and particularly interceptors placed in space, it may cause Russia to lose its second-strike capabilitity.[9]

Even if its missile deterrent is not threatened, Russia will no doubt, and especially now with the ABM treaty gone, develop methods for its warheads to penetrate the NMD system. The Russian reactions have been extremely critical.[10] For Russia the withdrawal from the ABM treaty and the building of a national missile defence system is not only a question of its missile forces and international stability, but also about technology. For Russia a most sensitive point is the technological gap and the fact that Russia is losing out after, during the Cold War, having maintained technological parity in many fields.[11]

The early US missile defence programmes were also directed against the Chinese threat. Particularly in the 1970s, China was expected to acquire strategic nuclear weapons and to threaten the US. China has, in fact, acquired a nuclear force of some two dozen single-warhead silo-based missiles and submarines capable of carrying ballistic missiles. Even though the Chinese forces are not on alert, they are ready to launch at short notice. The Chinese deterrent today is a question of mere possession of nuclear weapons being enough to prevent a nuclear attack. The national missile defence being built is designed to defend the US against an attack by tens of missiles, which is the size of China's ICBM force. China's nuclear deterrent would accordingly lose its meaning. The United States government has tried to assure China that their plans for a missile defence shield is not a threat to Beijing. The US deputy Secretary of State, Richard Armitage, commented during consultations with the Chinese: 'We believe if we have a limited – limited – defense against a handful of missiles, that in no way eliminates China's strategic deterrent, so it is not a threat to China.'[12]

The Chinese, however, are not convinced by American assurances. Given the situation, China has two basic options. One is to develop countermeasures so that the Chinese nuclear force will be able to penetrate the US national missile architecture. The second is to increase the size of its nuclear force or to deploy more missiles and/or deploy multiple warheads on missiles. The Chinese have protested against the plans and claimed that they have a profound negative influence on the global and regional strategic balance and stability and will trigger a new round of the arms race.[13] And the Chinese are indeed in the process of modernizing their nuclear forces.

The current threat image is the one of rogue states. These states included until recently North Korea, Iran and Iraq, sometimes also Syria and Libya. They are possible terrorist havens and harbour both anti-US sentiments and terrorist networks. The assessments of these threats are not about why and how these states would attack the United States with ballistic missiles, but rather about access to the necessary technology. Since the late 1990s a number of assessments have been made in the US about the ballistic missile threat and the access to weapons of mass destruction technologies. The first report was by the Rumsfeld Commission in 1998,[14] which for the first time emphasized the potential of missile programmes in emerging missile states. The assessment was that nations such as North Korea, Iran and Iraq could inflict major destruction on the US within about five years of a decision to acquire ballistic missile capability (ten years for Iraq). National Intelligence Estimates (NIE) in 1999 and 2001 have followed up and updated these assessments. The NIE of 2001[15] estimated the extent of the future missile threat right after the 9/11 crisis.

According to this estimate, a potential threat of intercontinental ballistic missiles existed from three countries, North Korea, Iran and possibly also Iraq. The most advanced capabilities were estimated to be in North Korea, which had even carried out tests.[16] North Korea was expected not only to possess nuclear capabilities, but also to develop chemical and biological weapons. According to the estimate, Iran was expected to have intermediate ballistic missiles and a programme to develop intercontinental systems. Iran would be expected, according to the report, to be able to launch an ICBM missile by the end of the decade. Even though Iran, according to the report, did not have a nuclear weapon, it was active in the exploitation of civil nuclear technologies. A third country, Iraq, was according to the Intelligence Council's estimate not able to experiment with long-range missiles before 2015, although reservations were made for closer cooperation with North Korea. Iraq was expected to have development programmes for both biological and chemical weapons.

The current situation (October 2004) has to some extent confirmed these estimates. In Iraq no indications have been found of a large-scale missile development programme. The weapon inspection carried out just before the second Gulf War found missiles with a range longer than the 150 km approved by the United Nations. This distance was, however, only exceeded by some tens of kilometres. No nuclear, biological or chemical warfare programmes have been found, a fact that was confirmed by the Iraq Survey Group which presented its report in October 2004. Iran is negotiating the inspection of its nuclear facilities with the IAEA in order to confirm its non-military use of nuclear power. No agreements have been reached and there are indications that Iran will continue its nuclear programme with a potential for weapon-grade enriched uranium production.[17] Ballistic missile development has not been discussed in this context. North Korea, although a party to the Nuclear Non-Proliferation Treaty, has openly defied the treaty. It has expelled the IAEA inspectors and has, in spite of agreements to the contrary, continued to build factories potentially capable of developing nuclear weapons. Combined with its capacity to fire missiles, this presents a potential threat, if not to the US, then at least to South Korea and Japan. A pre-emptive strike as in Iraq is not on the agenda and multilateral negotiations are under way between North Korea and the US, China and Russia.

In addition to the current threat of rogue states there are two additional threat images. One is the threat of terrorists getting access to nuclear material and making a crude bomb. Here the NIE points out that the easiest way of taking a crude bomb to the US is not in a long-range ballistic missile. This assessment is also supported by the advice of 49 retired generals and admirals, who, in advising Bush to shelve his missile defence start-up plan, urged him to spend the money instead on anti-terrorist defences at US ports, borders and nuclear weapon depots.[18]

Secondly there is the risk of accidental launches. This risk comes from two states only, China and Russia. Accidental launches from rogue states or other states with emerging missile capabilities are of course a possibility, but such launches being able to target the US are improbable and would not be a motivation to invest billions of dollars in a missile defence system. A Chinese accidental launch is not probable, because the warheads and fuel are stored separately from the missiles.[19] A Russian accidental, unauthorized or erroneous launch is a potential threat, but also one on a scale where the missile shield would not give protection. The Ballistic Missile Organization in a report to the House National Security Committee points out that a Russian unauthorized attack would deploy some 60–200 warheads, as Russian ICBMs in a division are interconnected.[20] Thus no stray missile could be accidentally launched.

In sum, throughout the previous paragraphs we have seen that the issue of missile defence has been on the political agenda for much of the last 50 years. The threats have oscillated, and so accordingly has the mission of the system. But the solution and response to the threats have throughout most of the period been a technological fix through missile defence. This can especially be seen after the Cold War. The threats have changed but the means have stayed the same. This is the case even though ample arguments have been provided, questioning whether missile defence has been up to its various missions. In the following I shall try to provide an alternative and complementary explanation as to why missile defence has continued to be seen as the solution.

The warfighter's edge

US military operations and the process of transformation called the 'revolution in military affairs'[21] are dependent on assets in space. Two US documents explore these vulnerabilities in space and develop the US policy in space. The most important is the report of the Commission Assess United States National Security and Space Management ai Organization, fathered by the present US Secretary of Defense, Donal Rumsfeld, in 2001, here called the Rumsfeld Report. This report exposes how space becomes a part of US security policy.[22] The second is the United States Space Command's *Vision for 2020*.[23]

Space is the top US national security priority, according to the Rumsfeld Report: 'The present extent of US dependence on space, the rapid pace at which this dependence is increasing and the vulnerabilities it creates, all demand that US national security space interest be recognized as top-national security priority.'[24]

The US government, in particular the Department of Defense and the intelligence community, are not yet, according to the report, focused to meet the national security space needs of the twenty-first century. The report concludes 'that a number of disparate space activities should promptly be merged, chains of command adjusted, lines of communication opened and policies modified to achieve greater responsibility and accountability'.[25]

The report accordingly outlines organizational changes for the development and deployment of space capabilities for US security interests in space. Investments in science and technology are needed, as are people in order for the US to remain the world's leading space-faring nation. The government has to sustain its investment in breakthrough technologies in order to maintain its leadership in space. The report concludes further:

> We know from history that every medium – air, land and sea – has seen conflict. Reality indicates that space will be no different. Given this virtual certainty, the US must develop the means both to deter and to defend against hostile acts in and from space. This will require superior space capabilities. Thus far, the broad outline of US national space policy is sound, but the US has not yet taken the steps necessary to develop the needed capabilities and to maintain and ensure continuing superiority.[26]

The report sees space as a medium much the same as air, land or sea. The US will in the future conduct operations to, from, in and through space in support of its national interests both on earth and in space. Furthermore, space systems will transform the conduct of future military operations.

The United States Space Command's *Vision for 2020* develops the necessary operational concepts for space, the warfighter's edge. These are full-spectrum dominance and conduct of joint operations. Full-spectrum dominance, i.e. to defeat any adversary and control any situation across the full range of military operations, rests on information superiority, which in turn implies innovation and new ideas. The conduct of joint operations requires interoperability for effective joint, multinational and inter-agency operations and the coordination of information assets, collaborative planning in crisis, and compatible processes and procedures. All this, in turn, requires speed and overwhelming operational tempo, precision engagement, and joint command and control. These operational concepts can only be obtained, according to the vision, through control of space. In short, the conclusions reached in these documents are that space becomes a key area of responsibility. Space is important to secure American national interests, accordingly securing space in itself becomes in turn a vital national interest. In this vision it is underlined that global surveillance is also the key to national missile defence.[27]

The American Philosophical Society arranged, in April 2000, a symposium on 'Ballistic Missile Defense, Space, and the Danger of Nuclear War'. At this meeting, strong arguments were put forth for prohibiting the deployment of weapons in space.[28] On the other hand the point was made that 'almost any satellite can be considered a weapon, and that the presence in space of weapons by some reasonable definition is inevitable'.[29] The weaponization of space and the potential effects of missile defence on it is therefore a question of whether any distinction at all can be made between the militarization of space (military use of information and communication systems) and the actual deployment of offensive weapon systems in space.

Research to clarify these questions is scarce. An interesting exception is the anthology collected by James Moltz called 'Future Security in Space: Commercial, Military and Arm's Control Trade-offs'. As space is already militarized, the authors point out that the next step, the weaponization of space, represents a new qualitative challenge. As noted above, this challenge is one of definition. But it is also one of content and legitimacy. On the one hand, weaponization of space is seen as a legitimate interest. One of the authors, Stephen Lambakiss, claims that there is a strong case for possible uses of space in order to enhance ballistic missile defence and in order to achieve a more effective space control. He concludes that: 'In the end all thoughts of the weaponization of space have to be seen in relation to rightful needs for defense in a world characterized by the spread of weapon technologies and by political strategic surprises.'[30] Other authors underline the importance of the common collective responsibility of the international community in limiting the weaponization of space. The risk of conflict and wars increases when every country able to launch a long-range missile, or deploy a satellite, may place 'weapons' in space.

Weaponization of space and the defence of space-based assets are linked to missile defence through global surveillance and space-based interceptors. First, space-based infra-red sensors are used in all types of interception, boost phase, midcourse and terminal. These eyes of the system need protection, a fact that will provide the necessary arguments and funding for the development of weapon systems to protect them for example against anti-satellite weapons. Second, space-based interceptors, in turn, are at the core of the boost-phase defence. Interceptors from space will be able to target the launched missile within seconds of the launch, rendering the defence more effective than the more problematic midcourse interception. This way the problem with countermeasures may be overcome. The two options for space-based missile defence, directed and kinetic energy weapons, encounter technical difficulties today. Once developed and deployed, they will, however, greatly impact the weaponization of space and provide highly offensive capabilities.[31]

Missile defence and securing American national interests in space are thus entwined. On the one hand the space-based sensors of the projected missile defence system strengthen US dependency on space, thus further explicating the need for the defence of space assets. On the other hand, missile defence technologies promise to provide that protection. The future technology necessary for space-based interceptor systems is akin to that needed to provide both protection against killer satellites as well as offensive ASAT capabilities. The mission is defence, the potential offence.

Missile defence is thus linked to the potential threats to US interests in space. However, this link is reinforced by the role of the missile defence project in developing new technology – the second part of my argument in explaining missile defence development.

Redesigning the technology base and the return of spin-off

New technology is needed for space control and surveillance. The threat from the rogue states and terrorist regimes provides the arguments for increased military R&D budgets. Indirectly this is simultaneously a way to build a competitive edge for the future in general. New competitive technologies, for example laser and sensor systems, will be developed, which will support the general US technological lead in the future. The interesting question is: why does the US government have to go about this by the way of military innovations and technology?

In the US economic system, the government should not pick winners and losers. This is an often-heard statement against state involvement in industrial and technological policy. Unlike European governments, which have paid for industrial and technological developments through grants, loans and subsidies to their enterprises, this is not the case in the US. The government should not pick the winners. This is the task of the marketplace in the US, with one exception – the military. The military has always been a way for the US government to support its national technological and industrial base.[32] The technological race inherent in the Cold War was an extremely good platform to channel extensive resources for technological developments into large civilian or military companies developing technology for military uses. The effort was extensive and sold to the public as the spin-off paradigm. According to this, resources spent on military technologies would profit the commercial sector as well, through spin-off. A number of technological developments in computer science were used as examples to show the advantages of this way of supporting commercial technologies and the American lead in technology.

The end of the Cold War, however, left the world of technological policy in the US in a state of turmoil. Large expenditures on military R&D as well as on military industrial contracts could no longer be supported as the technological race had collapsed. The decline in the Department of Defense procurement budgets left the US weapon industry shrinking to align supply to fit demand. A number of companies left the military field; others tried to gain strength through mergers and acquisitions and expanding the export market.[33] The Office of Technology Assessment of the US Congress produced alarming reports on the need for the military industry to restructure, with titles such as 'Redesigning Defense: Planning the Transition to the Future US Defense Industrial Base', 'Adjusting to a New Security Environment: The Defense Technology and Industrial Base Challenge' and 'After the Cold War: Living with Lower Defense Spending'.[34] These studies made one point extremely clear. There was a need for the defence industry to merge with the broader commercial technology and industrial base. It would be necessary for the military industry to take advantage of commercial technology developments and to make changes in the US procurement system in order to allow for this.

Worst-case scenarios were dramatic:

> Should defense prove unable to work with the commercial sector, it would be forced to retreat into a specialized ghetto, with military technology focused on specialized requirements pursued by a few private firms or by dedicated government arsenals. With little access to commercial technologies, components and sub-systems, costs would rise and capabilities shrink. In this extreme, this scenario ends in collapse – a debilitated and isolated enclave.[35]

Furthermore, in the early 1990s a number of studies were published criticizing the spin-off ideology. In a famous study, *Beyond Spin-off*, a number of highly regarded scientists at the Harvard Business School[36] showed that the spin-off was highly overvalued. Through examples they showed that the extent of the phenomenon was much less than expected and that supporting commercial technologies through military innovations was not only expensive but also disrupted the emergence of competitive commercial technologies. Other parallel studies[37] demonstrated that in many fields commercial technologies developed in commercial industries close to the competitive marketplace were cheaper and more advanced than their military counterparts. This was also true in computer science where the military computers were much slower and more cumbersome than the technologies used in commercial computers and communication technologies. The conclusion on the spin-off paradigm was a radical questioning of the previous policies:

> The important point about the spin-off paradigm is not that it was a half truth at best, but that the unusual circumstances of the Cold War world did not force the Americans to question it. Few asked whether the spin-off was an efficient way to link government efforts to commercial performance.[38]

Given this situation, the Clinton administration in the mid-1990s decided on a new paradigm for technological development, the *dual-use concept*. Here the state would support research and development work on technologies with both potential military and civilian uses. For example, sensor technology could be developed on a generic level before the military and civilian uses diverged. This would give the state a possibility for funding research work, which at a later stage would have military applications. A new technological partnership administered by the Pentagon, the so-called Technology Reinvestment Project, was implemented between commercial and military companies, which now would work together in the early pre-competitive stages of the research process.[39]

This implied a drastic change in the American way of working. The Department of Defense, which had earlier controlled all specifications for military material, was increasingly expected to use commercial specifications. The military would get access to advanced commercial technologies and commercial firms could benefit from early participation in technological development. Instead of spin-off, which meant a longer transformation process of military innovations into the civilian, synergies could be identified at an early stage and resources would thus be used more effectively, or so the argument goes.

The dual-use concept was not a success. It was not easy to introduce commercial specifications into the military procurement process. Civilian and military technologies diverged as to their requirements already in the research phases. Secrecy of military developments was a further hindrance. Military research laboratories have special ways of working, not immediately adaptable to combined civilian and military work. By the end of the 1990s, considerable political pressures existed for increasing military research and development. Missile defence provided an ideal case. Republicans have always advocated high expenditures on military research and development, advocating at the same time state non-involvement in industrial policy. The military industry was yearning for high-skill contracts on military research. The Bush administration's decision on missile defence deployment opened the way for research into guidance systems, advanced information and communication systems, technologies aimed at detecting and tracking missiles as well as space-based lasers. Innovations in these fields are not only useful in missile defences but also in developing new warfare and global surveillance.

The R&D budget of the military sector after the Cold War was at an all-time low. In 1991 it amounted to $35 billion. Today, the budget proposal for 2005 is almost double, $67 billion. Here missile defence plays a large role. James Lindsay and Michael O'Hanlon[40] have analysed the defence budget in this light. Bush's defence budget for 2003 included a strong support to the 'midcourse defence', which means that the enemies' attacking missiles are destroyed outside the atmosphere, in space. The budget includes research on a space-based laser and interceptor system, which not only may be used in midcourse defence, but also in the boost phase, i.e. before the missile leaves the atmosphere.

In conclusion, the plan to build a national missile defence can be seen as a way of 'back to normalcy'. There is again a large national project, which enables the US government to invest in military technology. It is a project to defend the American people in all the 50 states after the events of 9/11. Bush's goal, 'to protect our people with our technology', is being heard and reflected in the budgets. Technological supremacy is in the future maintained, not by the fussy dual-use concept or by merging the commercial and military industrial bases, but by increased R&D budgets and investments in military technology. The military industrial sector has regained its previous strength.

Even the spin-off paradigm is being revived in discussions of the potential for the commercial use of space. A number of authors in the Moltz anthology stress the importance of commercial space programmes and the possible lead that commercial space uses would take in relation to military use. Commercial space programmes could in fact support US military interests in space. This would apply in particular to communication satellites and imagery. This spin-off effect is also underlined in the Rumsfeld Report, according to which space would be expected to revolutionize both commercial and social activity. Space will come into homes, enterprises, schools, hospitals and government offices. The technologies of space will be applied to traffic, health, environment and agriculture. In short, space will not only be the warfighter's, but also the salesman's edge.

Against all treaties

The importance of the twin goals of space control and unconstrained technological development become clearly visible when looking at how the US deals with international treaties, especially the ABM treaty and treaties limiting the use of outer space for military purposes. On the one hand the US is extremely concerned over nuclear proliferation, particularly to rogue states and terrorist groups. On the other its own policies undermine international conventions and arms control agreements especially with Russia, the country with the largest nuclear arsenal and one under decay.

Given the US decision to build a national missile defence, the ABM treaty was on the way. The ABM treaty not only specified quantitative and qualitative goals for missile defences, it also limited technological development. Article V states that each party undertakes not to develop, test or deploy ABM systems or components, which are sea-based, air-based, space-based or mobile land-based. The Bush administration has been arguing that the treaty is thwarting the development of missile defence technology. While it permits the testing of fixed land-based anti-ballistic missile systems, it prohibits all mobile (sea, air and space-based as well as mobile land-based) systems. Philip Coyle, senior advisor at the Center for Defense Information and former director of Operational Test and Evaluation at the Department of Defense, and John Rhinelander, former legal advisor to the US SALT I delegation that negotiated the ABM treaty, have argued that national missile defence is in its infant stages and that it will take a decade or more to mature:

> The bottom line is that the development of national missile defense could go on for many years without violating the ABM Treaty – leaving time for negotiations, if needed, about future changes to accommodate whatever NMD system proves most promising. At the same time, it must be stressed that it is relatively easy to design a national missile defense testing program with the sole goal of busting the treaty. The current testing plans seem bent on that course. But doing so would run major risks, both technical and political.[41]

The authors also underline that US withdrawal from the treaty could lead to a collapse of the larger arms control regime. Lindsay and O'Hanlon agree with this conclusion, and propose, for the time being, that the US forgoes testing and deployment of missile defence weapons in space.[42]

The Russians have indicated a willingness to modify the treaty. The Foreign Minister of the Russian Federation asked in an article in *Foreign Affairs*: 'Does there exist, in principle, the possibility of creating a national antiballistic missile system while preserving the 1972 ABM treaty as a cornerstone of strategic stability?'[43]According to him, Russia is no less interested in finding a solution to the ballistic missile threat and a solution should be sought together, or at least not to the detriment of each other's interests. When Presidents Bush and Putin met in Crawford, Texas, in November 2001 there were rumours circulating that they would agree to modify the treaty. Washington would get greater freedom to test and deploy missile defence technologies and Moscow would get limits on the ultimate size and nature of any US defence. According to Lindsay and O'Hanlon the logic was straightforward. To maintain Russian cooperation in the war against terror, the US could accept constraints on ambitious defences, which in any case would not be ready for years.[44] As we know, no agreement was reached and the US withdrew from the treaty in December 2001. This underlines the importance of the research and development elements – especially associated with space – in the missile defence project. If the sole purpose was a limited missile defence, a solution would probably have been reached. It was not, however. The US wanted freedom to pursue a broad range of technologies.

Of further importance is the 1963 Limited Test Ban Treaty, which prohibits nuclear weapon explosions in outer space. The 1967 Outer Space Treaty prohibits placing weapons of mass destruction in space, on the moon or other celestial bodies, and using the moon and other celestial bodies for any military purposes. Additionally a number of arms control treaties prohibit the US and Russia from interfering with each other's use of satellites for monitoring treaty compliance. Finally, the 1980 Environmental Modification Convention prohibits all hostile actions that might cause long-lasting, severe or widespread environmental effects in space.[45]

The Rumsfeld Report on Security and Space Management points out that in these treaties there is no blanket prohibition in international law on placing or using weapons in space, applying force from space to earth, or conducting military operations in and through space. Furthermore, extending the principles of the UN Charter (Article III) to the outer space treaties provides for the right of individual and collective self-defence, including 'anticipatory self-defence'. The non-interference principle established by space law treaties would be suspended among belligerents during a state of hostilities.[46] These thoughts about space underline the American concern with protection of its space assets, and the importance the US attaches to maintaining its freedom of action.

The current US positions towards regulating the weaponization of outer space follow these lines. Both Russia and China have proposed at the UN Conference on Disarmament to expand the Outer Space Treaty to ban all types of weapons. The US has opposed this. In November 2000 the US was one of the three countries that refused to vote for a UN resolution on the need to prevent an arms race in space. John Bolton, the US Under Secretary of State for Arms Control and International Security, has underlined for the Conference on Disarmament that 'the current international regime regulating the use of space meets all our purposes. We see no need for new agreements.'[47] The Bush administration has not only taken the view that no new agreements are needed and that the US wants to exploit space for military purposes; it has also stated: 'A key objective . . . is not only to ensure US ability to exploit space for military purposes, but also as required to deny an adversary's ability to do so.'[48] In this, missile defence research and development plays an important role.

The seamless web

'The seamless web' is a concept used in technology theory, particularly by the constructivist school. It denotes that technology is intimately linked to society.[49] In fact, technology and society can no longer be seen as separate, they are intertwined. If the social organization around a certain technology collapses, so does the technology itself. Power is seen as the result, not the starting point for a certain technological development. Arguments and alliances matter in the process. Although we can talk about the missile shield as a technological trajectory, with stable financing through decades, it is not a technological imperative autonomous of the socio-political environment. The tasks of missile defence varied according to this environment.

Especially interesting in the current developments is the importance of one person. Presently Secretary of Defense, Donald Rumsfeld headed the commission which in 1998 made the ballistic missile threat explicit, a conclusion followed up by a number of National Intelligence Estimates. The US threat perception thus changed. He was also the person in charge of the commission defining US space interests and making space a top priority for US security. As the Secretary of Defense in the present Bush administration, he obviously played a central role in making the decision to actually deploy a missile defence system by the end of September 2004; he is no doubt the 'Spiderman' in the seamless web of missile defence.

Of course, there are alliances. A national missile defence is not only supported by Republicans. It was President Clinton who signed the National Missile Defense Act. The commission headed by Donald Rumsfeld on the ballistic missile threat included three Democrats and was thus bipartisan. The conclusions were unanimous. Reduced defence spending and lower R&D budgets for the military at the beginning of the 1990s obviously made new funding and thus missile defence interesting to large parts of the scientific community. Large military industries in charge of missile defence systems development such as Boeing and Raytheon are also strong allies in technology policy. Finally, the American people may support the spin-off argument once more, as the US lead role in commercial exploitation of space is no doubt in their national interest, and of course so is the potential of protection against the ballistic missiles of rogue states.

There are also critical voices. The State Department was openly critical of the Rumsfeld Report in 1998. The National Intelligence Estimates have referred to the State Department as an 'agency with the less alarmist view' on the ballistic missile threat from the rogue states, than that of the CIA or the Pentagon. The positive view of scientists at large military establishments is balanced by the more critical attitude of the concerned scientists as expressed for example in the 'Countermeasures' report. The main dissidents, however, come from the international community.

The Chinese and Russians have not only been opposed to the missile defence plans, they have worked for treaties to prevent the weaponization of space. All regional great powers such as the European Union, Russia, China and Japan are each preparing to decrease their vulnerability in technological development, particularly in relation to the US dominance in space. Here the European effort to build its own space-based global positioning system, Galileo, is an interesting case. Not the least, since Europe is teaming up with China in space efforts, which in no way, at least not yet, challenges US domination in space.[50]

To summarize, the argument of this chapter is not that the threat of rogue states and their potential access to ballistic missile technology is not there. What is remarkable is that the social dynamics of missile defence reach much wider in American society. In this wider picture, technological policy and the control of space make the difference. To understand the politics of the American missile defence plans these two elements have to be integrated as well as the threat from rogue states. It is the juxtaposing of all three elements that makes missile defence such a compelling policy choice. Further, the importance of securing space assets and the need to stimulate technological development in general contribute to explain why the missile defence architecture has this distinct appearance. To state it in a different way: Had the need to invest in military technology not been there, the potential missile threat from rogue states would conceivably have been met by other means such as sanctions or pre-emptive strikes. Had the missile defence plans not supported the defence of space-based information systems and the goal of global surveillance, there might have been less political support for the necessary R&D funding. Seen in this context, missile defence is more than just defence against external threats. It is also, and maybe foremost, a symbol of technological supremacy and a means to secure world leadership in technology in the future also.

Notes

1 For analysis of the National Missile Defence system, see James Lindsay and Michael O'Hanlon, *Defending America: The Case for Limited National Missile Defense*, Washington, DC: Brooking, 2001; Bruno Tertrais, *US Missile Defense*, London: Centre for European Reform, 2001; and Ingemar Dörfer, *Ballistic Missile Defense. Det amerikanska programmet – Säkerhetspolitiska konsekvenser*, Stockholm: Totalförsvarets Forskningsinstistut, 2002. Online. Available HTTP: ⟨http://www. foreignpolicy.com/www/board/watts/tlm⟩.

2 See Lindsay and O'Hanlon, op. cit. p. 6.

3 Ibid., p. 113, citing *Washington Post*, 24 May 2000.

4 Ibid., pp. 101, 110.
5 Union of Concerned Scientists/MIT Security Studies Program, *Countermeasures: A Technical Evaluation of the Operational Effectiveness of the Planned US National Missile Defense System*, Boston, 2000. (Authors: Andrew M. Sessler [Chair of the Study Group], John M. Cornwall, Bob Dientz, Steve Fetter, Sherman Frankel, Richard L. Garwin, Kurt Gottfried, Lisbeth Gronlund, George N. Lewis, Theodore A. Postol, David C. Wright.)
6 Ibid., p. xxi.
7 Technological trajectory is a concept introduced by Giovanni Dosi in the context of economic theory; see Dosi *et al.*, *Technological Change and Economic Theory*, New York: Pinter, 1988. The concept stresses the continuity of technological developments.
8 The social construction of technology focuses on the arguments and alliances in the process of technological development. See Wiebe Bijker and John Law, *Shaping Technology/Building Society*, Boston: MIT Press, 1992; Wiebe Bijker, Thomas P. Hughes and Trevor Pinch, *The Social Construction of Technological Systems*, Boston: MIT Press, 1987. (See especially the introductory essay by Bijker and Pinch.)
9 Mathematical scenarios for this are indicated in Altes F. Korthals, *An Analysis of the US Missile Defence Plans: Pros and Cons of Striving for Invulnerability*, The Hague: Advisory Council on International Affairs, 2002.
10 For Russian statements on NMD, see Union of Concerned Scientists/MIT Security Studies Program, op. cit., p. 109, and Igor Ivanov, 'The Missile-Defense Mistake', *Foreign Affairs*, 79(5), September–October 2000; Aleksej Arbatov, 'ABM-traktaten og terrorismen', *Udenrigs*, 1, 2002. Igor Ivanov was the Minister of Foreign Affairs of the Russian Federation and Aleksej Arbatov the Vice-Chairman of the Defence Committee of the State Duma.
11 For a review of Russian technology policy, see Tarja Cronberg, *Transforming Russia from Military to Peace Economy*, London: I.B. Taurus, 2003.
12 See CNN.com/World, 11 May 2001, and *People's Daily Online*, 3 September 2001.
13 See Union of Concerned Scientists/MIT Security Studies Program, op. cit., p. 111, for Chinese statements on NMD.
14 Report of the Commission to Assess the Ballistic Missile Threat to the United States (Rumsfeld Commission Report), 15 July 1998.
15 National Intelligence Council, *Foreign Missile Developments and the Ballistic Missile Threat Through 2015*, 2001. Online. Available HTTP: ⟨http://www.cia.gov/nic/pubs/other_products/unclassifiedballisticmissilefinal.htm⟩.
16 Lindsay and O'Hanlon, op. cit., p. 203.
17 See for example 'Europeans Decry Iran's Nuclear Plans', *Herald Tribune*, 2 April 2004, 'Iran Plans to Take Another Step Down Nuclear Path', *Herald Tribune*, 2 September 2004, and 'Iran Plans to Turn Uranium into Substance for Weapons', *Wall Street Journal Europe*, 2 September 2004.
18 *Herald Tribune*, 2 April 2004.
19 Union of Concerned Scientists/MIT Security Studies Program, op. cit., p. 9.
20 Ballistic Missile Defense Organization, *National Missile Defense Options*, prepared in response to a request from the House National Security Committee, 31 July 1995, p. 4.
21 Bertel Heurlin, Kristian Søby Kristensen, Mikkel Vedby Rasmussen and Sten Rynning (eds), *New Roles of Military Forces: Global and Local Implications of the Revolution in Military Affairs*, Copenhagen: Danish Institute for International Studies, 2003.

22 Report of the Commission to Assess United States National Security and Space Management and Organization, 11 January 2001 (here called the Rumsfeld Report).
23 United States Space Command, *Vision for 2020*, Washington, DC: US Government Printing Office, 2000.
24 Rumsfeld Report, op. cit., p. 99.
25 Ibid.
26 Ibid., p. 100.
27 United States Space Command, p. 37.
28 R. Garvin, 'Space Weapons or Space Arms Control', a paper presented at the American Philosophical Society Annual General Meeting in the symposium *Ballistic Missile Defense, Space, and the Danger of Nuclear War*, 2000.
29 David Finkelman's paper to the above meeting, p. 2.
30 James Moltz, 'Future Security in Space: Commercial, Military and Arms Control Trade-offs', Monterey: Center for Nonproliferation Studies. Occasional Paper No. 10, 2002.
31 For these technological links, see for example I. Safranchuk, 'The Link Between Missile Defense and Space Weaponization', *International Network of Engineers and Scientists Against Proliferation, Bulletin 20*. Online. Available HTTP: ⟨http://www.inesap.org/bulletin 20/bul20art05.htm⟩.
32 See for example Office of Technology Assessment/US Congress, *Redesigning Defense: Planning the Transition of the Future US Defense Industrial Base*, Washington, DC: OTA-ISC-500, 1992.
33 For the consolidation of the military sector in the US, see Tarja Cronberg, Anders Aero and Eric Seem, *Technological Powers in Transition: Defense Conversion in Russia and the US 1981–1985*, Copenhagen: Akademisk Forlag, 1996, p. 94.
34 Office of Technology Assessment, US Congress, *Redesigning Defense: Planning the Transition to the Future US Defense Industrial Base*, Washington, DC: US Government Printing Office, 1991; Office of Technology Assessment, US Congress, *Adjusting to a New Security Environment: The Defense Technology and Industial Base Challenge*, Washington, DC: US Government Printing Office, 1991; Office of Technology Assessment, US Congress, *After the Cold War: Living with Lower Defense Spending*, Washington, DC: US Government Printing Office, 1992.
35 John Alic, Lewis Branscomb, Harvey Brooks, Ashton Carter and Gerald Epstein, *Beyond Spin-off: Military and Commercial Technologies in a Changing World*, Boston: Harvard Business School Press, 1992, p. 363.
36 Ibid.
37 See for example J. Stowsky, 'Beating Our Plowshares into Double-Edged Swords: The Impact of Pentagon Policies on the Commercialization of Advanced Technologies', The Berkeley Roundtable on International Economy, Working Paper 17, 1986.
38 Alic *et al.*, op. cit., p. 113.
39 For a discussion, see Cronberg *et al.*, op. cit., and ARPA, The Technology Reinvestment Project, *Dual Use Innovation for a Stronger Defense*, Washington, DC, 1995.
40 James Lindsay and Michael O'Hanlon, 'Missile Defense after the ABM Treaty', *The Washington Quarterly*, 25(3), 2002, pp. 163–176.
41 Philip Coyle and John Rhinelander, 'National Missile Defense and the ABM Treaty: No Need to Wreck the Accord', *World Policy Journal*, 18(3), 2001, pp. 21–22.

42 Lindsay and O'Hanlon, op. cit., p. 175.
43 Ivanov, op. cit., p. 16.
44 Lindsay and O'Hanlon, op. cit., p. 165.
45 Quoted in the Rumsfeld Report, op. cit., p. 37.
46 Ibid.
47 Statement by John R. Bolton, US Under Secretary of State for Arms Control and International Security, to the Conference on Disarmament, Geneva, 24 January 2002.
48 2001 Quadrennial Defense Review (QDR), released 1 October 2001; quoted from T. Hitchens, *Weapons in Space: Silver Bullet or Russian Roulette? The Policy Implications of US Pursuit of Space-Based Weapons*, Center for Defense Information, 2002. Online. Available HTTP: ⟨http://www.cdi.org/missile-defense/spaceweapons.cfm⟩.
49 The term was first used by Thomas Hughes. See Bijker, Hughes and Pinch, op. cit.
50 See for example Istituto Affari Internazionali, *International Report on Space and Security Policy in Europe: Executive Summary*, Rome, 2003; and a Danish article, 'EU udfordrer USA I rummet', *Berlingske Tidende*, 31 October 2001.

3

MISSILE DEFENCE IN THE UNITED STATES

Bertel Heurlin

Introduction

The basic arguments of this chapter are, first, that the current US missile defence is based upon former experiences with missile defence; second, that missile defence closely associated with weapons of mass destruction has gained highest priority in American national security policy due to the 9/11 attacks; and third, that the superior argument for establishing an American missile defence is to maintain global long-term political-strategic superiority.

The chapter is structured in the following way. After first introducing the state of affairs for missile defence in the United States, second, five historical stages will be identified and analysed, using the long-term political-strategic objectives as a point of departure. Third, the national and international debate on missile defence is outlined and discussed. Fourth, the recent missile defence policy is explained according to a neorealistic perspective. Finally, a brief conclusion is presented.

Recent missile defence efforts

First we have to set the scene. The political justification for establishing the new American missile defence system is the emergence of new WMD threats from rogue states and rogue non-state actors.[1] The United States is obliged to reduce its vulnerability. The system is just a beginning. It is argued that it is necessary to commence setting up a system able to respond to rapidly changing threats exploiting advances in technology. On 26 August 2004, Secretary of Defense Donald Rumsfeld stated that:

> By the end of this year [2004] we expect to have a limited operational capability against incoming ballistic missiles. This represents,

> in my view, a victory for hope and vision over scepticism.... Rather than waiting years, sometime decades, for a fixed and final architecture, as has been the norm with many weapons systems, we will be deploying an initial set of capabilities that will evolve over time as technologies evolve over time.[2]

The American plans for missile defence for 2004 and 2005 are as follows. Six ground-based interceptors (GBIs) will be deployed in autumn 2004, followed by ten more by the end of 2005 at Fort Greely, Alaska. Further, four GBIs will be deployed in 2004 at Vandenberg Air Force Base, California. According to the initial planning schedule of the administration, 20 sea-based interceptors placed on three Aegis-class cruisers or destroyers as well as an unspecified number of Patriot PAC-3 interceptors are foreseen to be deployed in the near future. The GBIs are intended to attack incoming intercontinental ballistic missiles (ICBMs), and the sea-based and Patriot interceptors are expected to deal with short- and medium-range missiles. In order to take care of target detection and to receive tracking information, it is necessary to establish a connection between the existing system of infrared early warning satellites and an upgraded Cobra Dane radar at Shemya, Alaska, a new sea-based X-band radar, SPY-1 radars based on Aegis-class ships, and upgraded early warning radars in the UK in Fylingdales and in Greenland at the Thule base.[3]

The American missile defence is multilayered in time and scope. Three time layers could be identified: the instant effort, the medium term and the long term. The instant efforts started as the Bush administration in December 2001 declared that it would withdraw six months later from the 1972 ABM treaty in accordance with Article 15 of the treaty. A year later, on 17 December 2002, President Bush announced a firm date, September/October 2004, for commencing deployment of the missile defence system. The Bush administration's Nuclear Posture Review had earlier specified the deployment of an emergency missile defence capability sometime between 2003 and 2008.[4] The current system is called the Initial Defensive Capability (IDC), laying the foundation of the long-term integrated, layered Ballistic Missile Defence System (BMDS). The system has the effect of complicating the adversaries' efforts and reduces the military utility of ballistic missiles, discouraging the proliferation of such technology, as well as providing an effective deterrent.[5]

Regarding the medium-term development, the Missile Defense Agency operates with six blocks, 2004, 2006, 2008, 2010, 2012 and 2014. These blocks mark the evolutionary, spiral development, having to do with adding new capabilities based on technical maturity, upgrading existing capabilities, inserting technology, evolving requirements, procuring additional force, enhancing capability and extending the missile defence systems to allies and friends when appropriate.[6] In the long-term development the block approach will yield a fully integrated and layered BMDS, capable of defeating ballistic missiles of all ranges and in all phases of flight.

The multilayered pattern in scope refers to the three main phases in missile defence: boost phase, midcourse phase and terminal course. The reason for fielding a layered defence system and in this way being able to attack missiles in all phases of flight is to exploit opportunities to increase the effectiveness of missile defences and to complicate an aggressor's plans.[7]

Missile defence: developments and stages

In order to comprehend the current and future development it is necessary briefly to take a critical look at the short history of missile defence. It reveals a mixed picture. But since the development of ballistic missiles, the construction of a defence system has been part of the central strategic and political agenda. The political desirability and the technological possibility have been assessed differently in the five stages. Nevertheless, the missile defence programmes were never questioned as options and were never close to being cancelled. The R&D part continued uncontested during the years. This policy could be seen as an insurance premium for the United States technologically to be kept up to date regarding missile defence. Basically missile defence was promoted – or restrained – according to long-term political-strategic considerations. The intention in this section is to demonstrate how missile defence has all the time been used as a tool – and a very valuable one – in a superior political-strategic game as part of US security policy.

Five historical stages can be identified, analysed and characterized as follows:

1 From the 1950s to 1972: Missile defence is in the understanding of the US government considered politically fairly desirable and technically fairly possible.
2 From 1972 to 1983: Missile defence is considered politically hardly desirable and technically hardly possible.
3 From 1983 to 1989: Missile defence is considered politically highly desirable and technically fairly possible.
4 From 1989 to 2000: Missile defence is considered politically fairly desirable and technically fairly possible.
5 From 2000 onwards: Missile defence is considered politically highly desirable and technically highly possible.

This simple scheme denotes a specific political-technological relationship: partly that MD is a genuine long-term technological project, and partly that MD is a programme having the aim of signalling military and technological superiority and emphasizing MD as a 'political weapon', meaning a weapon not necessarily produced to be used in battle but rather used as part of a demonstration policy or a dissuasion strategy.

The first stage: 1956–72

The American missile defence project was initiated in 1956. The US Army began developing the Nike Zeus missile, based upon the air defence rockets Nike Ajax and Nike Hercules. From 1963 the Nike-X project took over, aiming at developing partly a super-fast ABM missile, the two-stage rocket Sprint, and partly a bigger and slower missile, Spartan, using the Nike Zeus as the point of departure. Spartan was expected to be equipped with a one-megaton nuclear warhead.[8] It was, however, striking that all the missile defence systems remained at the R&D stage. The regular process was that one system after the other was initiated, researched, developed, introduced, tested – but never deployed. Missile defence is in the first stage characterized as politically fairly desirable despite the fact that a defence system never materialized. The argument is that the so-called 'balance of terror', the situation where superiority cannot be established, was not fully politically accepted. It was necessary in general terms also to aim at protection, as the balance of terror was still fragile and delicate.[9] Using the term 'fairly desirable' it is indicated that missile defence did not have the highest priority. The *offence* as part of the 'assured destruction' strategy had. The assessment 'technically fairly possible' is justified by the fact that serious development programmes were pursued, programmes that *could* be implemented to operative systems, but did not materialize during this first stage. Missile defence was an option, taken seriously by the US, owing to its impact on long-term political-strategic considerations.

The second stage: 1972–83

The characterization of this stage is that missile defence politically was hardly desirable and technically hardly possible. What happened from 1972 to 1983 was that missile defence continued on a low flame. In 1972 a comprehensive US–Soviet arms control agreement was signed.[10] There were two parts: the SALT treaty, limiting the strategic offensive weapons and their warheads, and the ABM treaty, reducing the number of ABMs to 2 × 100 missiles to protect the capital and one ICBM base. The treaties were a demonstration of a close cooperation, a marked manifestation of the policy of détente between the superpowers. It was also, however, an indication of a continued confrontation: the United States as well as the Soviet Union declared that the strategic competition would go on in areas not covered by the treaties. In 1974 the parties went even further in the ABM issue. Only two years after the first ABM treaty the parties agreed that the maximum number of ABM missiles allowed should be reduced to 100, to be deployed around either the national capital or an ICBM base. The USSR chose to maintain the ABM sites around Moscow, the so-called Galosh system. The United States, however, had severe political-technological problems. An apparently effective system, Sentinel, turned out to be troublesome. After President Nixon took over the administration in 1969 the Sentinel system was expanded and renamed Safeguard. The system was tested and finally deployed in 1976 around the ICBM base in Grand Forks. It did not, however, last long. A few months later it was deactivated. The United States had chosen not to exploit the possibility in the 1974 treaty to match the Soviet Union in ABM deployment. Politically and technologically the system did not in the end prove cost-effective.

To characterize the American missile defence policy at this stage as politically hardly desirable and technically hardly possible is of course to simplify. It was, however, a fact that the United States after 20 years of missile defence development had recognized that no fundamental technological breakthrough had appeared, and that the offensive weapons in the nuclear age were so powerful and superior that defensive weapon systems would for some time remain in the background, in 'the defensive'. For the time being missile defence was not capable of matching the offensive capabilities. But, for political and military-technology reasons, R&D had to continue. Missile defence was, despite a temporary détente policy and technological shortcomings, still an option in the American long-term political-strategic considerations.

The third stage: 1983–9

1983 is the year of transformation in US missile defence. The Reagan administration took over in 1981. The military budgets went up, a tendency already introduced during the last years of the Carter administration. The policy towards the Soviet Union became increasingly confrontational. The stage from 1983 to 1989 can in terms of missile defence be typified as politically highly desirable and technologically fairly possible. How to explain this essential change? What happened?

In April 1983, President Reagan in an official statement introduced the SDI, the Strategic Defence Initiative.[11] SDI was a virtual bomb under the traditional strategic concept.[12] The result was a political explosion. Practically and operationally, however, nothing really happened. No missile defence was deployed. SDI was, as the term indicates, an initiative, the beginning of a comprehensive R&D programme, and the beginning of a new way of thinking, preparing for a fundamental change in the national security strategy of the United States. Signalling a new beginning was the main purpose. SDI implied a bottom-up concept. Under attack was the notion of the offensive as the crucial factor in the nuclear age, leaving defence as subordinated. The logic of the MAD strategy, mutual assured destruction, should be abandoned.

President Reagan wanted to get rid of this logic. He wanted to skip deterrence. The offensive threat to the adversary should not be the basic kind of defence. Defence was supposed to be a real defence: a protective shield over the United States, preventing incoming missiles from reaching their targets. The United States should regain its invulnerability. But abandoning deterrence in this context would also have severe implications for nuclear weapons. This also was a vital part of the message introducing SDI. Nuclear weapons should become 'obsolete and impotent'. Reagan's vision was, as expressed in his Star Wars presentation, 'that free people should live safe trusting that their security does not depend on the need for immediately to retaliate a missile attack, but that we could trace and destroy ballistic missiles before they reach our territory or that of our allies. Would it not be better to save lives than to revenge them?'

SDI was not just a bottom-up strategy. It was also a farewell to strategic balance between the superpowers. The vision of an invulnerable United States implied a United States emerging as the superior party vis-à-vis the Soviet Union. The United States seemed to be on its way to introducing a post-nuclear strategic world order, signifying the United States as a superpower not only having an overwhelming nuclear overkill capacity but also being in the possession of an effective anti-nuclear-weapons system. With this vision in place, the United States could in the long run control and organize the international system. This was, in reality, the essence of the 1983 message from President Reagan to the domestic security policy community, to the American population, to the Soviet Union and to the rest of the world. SDI was, as we have seen, no less than a revolution. It took, however, a long time to get the SDI revolution implemented. But to the Soviet Union SDI had serious implications. It marked the beginning of the end of the Soviet Union as an empire and a superpower. SDI can be assessed as the last decisive move in the virtual war between the two superpowers, ending with the voluntary Soviet surrender in the Cold War.

So, missile defence in the third phase was no doubt politically highly desirable. It was a political move, rocking the boat of nuclear deterrence. Also it can be assessed as technologically fairly possible. How come that there is this change from the former period where the characterization was technologically hardly possible? The answer is to be found in the SDI as a vision and the launching of the vision as a realistic concept. SDI was not about using or applying traditional and well-known technologies. They had already been tested, deployed and abandoned in 1976. The objective was a long-term R&D programme along the lines of the Manhattan atomic bomb project during the Second World War, experimenting with and using completely new technologies. Massive investments in innovation, research and development were supposed to open up the ultimate defensive system – or system of systems, diversified, multi-faceted, covering all kinds of defensive necessities, demonstrating the marked American technological superiority. An important precondition was a missile defence system based upon kinetic energy. The incoming missiles should be destroyed by physical collision or by laser beams. Using nuclear-tipped anti-missiles was obsolete and dangerous. The basic argument for SDI was that the technology was not at hand, not available: the creation of relevant new technology was required. The precondition was a determined scientific, economic and technological effort.

In conclusion, missile defence was not just a mere option; it had become an important part of the US long-term political-strategic considerations.

The fourth stage: 1990–2000

This stage is dominated by the structural sea changes in the international system: from bipolarity to unipolarity. The Soviet Union declined as a superpower. Missile defence had been heavily involved in this change.

The missile defence phase beginning with the end of the Cold War and ending with the beginning of the Bush administration is characterized by the notion that missile defence is politically fairly desirable and technologically fairly possible. 1989 had demonstrated the political justification of SDI. The Soviet Union had changed its national strategy fundamentally. The Soviet term was 'new thinking'. Part of the new thinking was to come to terms with the United States on the arms control issues. The USSR retreated from the arms race. It gave in and accepted practically all US suggestions in the negotiations, not least concerning verification and control. The Soviet Union began to liquidate its empire: globally by leaving Vietnam and Afghanistan, regionally by withdrawing from Eastern Europe, introducing what Gorbachev called the Sinatra doctrine ('I did it my way') to replace the Brezhnev doctrine, and finally locally by dissolving the union, as Russia declared itself a sovereign federal state.

SDI played, as indicated, a crucial role in these events. One may even claim that the Soviet Union surrendered voluntarily in the virtual war with the United States mainly due to the SDI. The Soviet military, political and economic establishment assessed SDI as the ultimate American step in the direction of establishing nuclear superiority. If you could not avoid this development, if you could not beat the United States, better join them. This is what more or less happened. The USSR accepted the Western international norms as global norms, and reduced itself to Russia proper. The end result was a partnership with the US, including NATO.

To the United States the period after 1989 was a sweeping transformation. A new world order had emerged. The enemy superpower had stepped down, leaving the United States as the only one on the podium. Using the nuclear weapons notions, the new situation can be characterized in the following way. With the Soviet Union stepping down as a superpower, followed by its reduction to Russia, a regional power with limited capabilities, the virtual war was over. The trump was SDI, certainly the most virtual of all the efforts, almost a cartoon weapons system in the virtual war.

But, factually, what happened to missile defence? With the end of the Cold War the United States could oversee the international horizon and state that the overwhelming threat against the United States had gone. Internationally, the security situation had never been better – no expectations of a conventional or a nuclear attack on the territory of the United States. Russia was, as mentioned, still a strong nuclear power. Increasingly Russia was considered a partner. Russia was no longer the USSR, but a kind of anti-USSR: it was democratic, it became a market economy, and it adhered to the new global norms of human rights and personal freedom. The vital threats to the United States could be identified as terrorism, with catastrophic terrorism at the top of the scale. Unintended accidental attacks from Russia and the possible rise of China to a top nuclear power with sophisticated nuclear systems were included in the scenarios. Also, actual and possible threats from rogue states emphasizing not only nuclear weapons, but any kind of weapon of mass destruction, were on the list.

Missile defence was a part of the answer to these threats. But only a part. The R&D fraction of the SDI continued. The ambitious SDI project was transformed in 1993 into a more directly applicable and narrower project. In principle, MD was politically fairly desirable. The budget did not change considerably. Four conditions were on the agenda: the state of technology, the effect upon allies and partners, relations with the states threatening the United States, and finally the costs of missile defence.

In 1998 a bipartisan commission chaired by Donald Rumsfeld recommended that the United States should develop a defence system against ballistic missiles. An overwhelming majority in the Congress (97–3 in the Senate and 317–105 in the House of Representatives) voted to deploy a national missile defence system as soon as it was technically feasible. A time schedule was introduced for its development. In 2000 a missile act was signed stating that the country should establish a missile defence, without, however, deciding how, where and when. On 1 September 2000, President Clinton announced that he would transfer the final decision for National Missile Defence (NMD) to his successor. Referring to the four criteria introduced earlier he stated that, first, the technology was not yet ready, second, that the impact on allies, partners and crucial states like Russia and China had to be studied more closely, third, that the confrontations with states threatening the United States had to be further reduced using diplomacy. The fourth criterion, the costs of NMD, was not mentioned.[13]

All in all, in the fourth stage missile defence was politically fairly desirable. It had a certain priority, but it was considered a project to be further investigated as to time and scope. The political drive from the SDI had diminished. Also, the relations to the technological possibilities were to be discussed. In general it was assessed that MD was fairly possible technologically, depending on the resources available. In conclusion, missile defence at this stage was still an important option in the American long-term political-strategic considerations.

The fifth stage: 2001 onwards

Would it be more correct to place the beginning of the fifth stage in 1998 with the Rumsfeld Commission and the general bipartisan acceptance of the need for a missile defence? Some arguments could justify this choice. What is important, however, is that the fundamental political missile defence decisions were not taken until the new Bush administration was in place. During the presidential campaign, Bush versus Al Gore, missile defence played a subordinate role. The Bush team, however, had missile defence pretty high on the agenda. The limitations of the ABM treaty on developing an effective national missile defence were exhibited as politically unacceptable. This policy was entirely in line with the general foreign policy line of the Republicans: to get rid of obsolete arms control agreements to ensure an American unilateralism not confined by existing or future international agreements that were not in line with vital American interests. Bush was thus in contrast to Clinton's attempts to make an American decision on MD dependent on a revised ABM treaty. The treaty had to disappear.[14] In summer 2001, half a year after the start of the new administration, three traditional questions were officially raised by the new administration to form the basis for assessing the MD. They certainly were not new. Does MD work? Is it cost-effective? And finally, how does it fit into the priorities of American security policy?

These criteria were put into perspective following the 9/11 attacks. Why support a system not technologically fully reliable, not fully cost-effective as it was supposed to protect against threats not specifically persuasive, not able to meet the threats coming from global terror networks using traditional, conventional and easily attainable technologies? Nobody expected al-Qaeda to launch ballistic missiles equipped with nuclear warheads, for the simple reason: why should they? For a non-territorial, non-state international actor there were smarter and simpler ways to terrify and to attack the United States. None the less, 9/11 was a decisive event as concerns MD. The 9/11 attacks demonstrated the extreme vulnerability of American society and the need to meet new threats on a number of levels. 9/11 proved that if the US was to be able to respond to terrorist attacks from states or non-states, an ability to defend itself from all kinds of attacks was required. An operative, comprehensive, coherent strategy of national security had to be formulated and implemented. This need was met in September 2002 when the new National Security Strategy was published.

Examining missile defence after the beginning of the Bush administration in 2001 has left no doubt about characterizing US policy as politically desirable. The project has full support from government, Congress and the population. Missile defence is high on the political agenda. The fact that the programme is already operational and that a comprehensive programme is planned and on track also supports the characterization of missile defence as technologically possible. There seems to be not only an expectation but also a cognition that the system is functioning.

Five stages in the American missile defence programme have been identified and analysed according to political and strategic-technological factors. In conclusion, during all five stages missile defence in differing forms and shapes has been at least an option as part of the American general political-strategic considerations – in two stages, 1983–9 and from 2000 onwards, even a vision.

The domestic and international debate: the pros and cons

The domestic debate in the United States on missile defence has in some instances been intense despite its common and general acceptance in Congress and in the opinion polls. This debate has primarily taken place among the elite belonging to the political, scientific and strategic policy community. In other instances the debate has been more or less absent, except for negative attitudes and demonstrations during circumstances where local populations were confronted with deployment of ABM systems around their densely populated areas.

The strong and heavy argumentation from the united front of opponents of a missile defence system can be condensed into the following simple assertions. First, missile defence has no *strategic* sense: it weakens deterrence, it weakens security, it is unreliable and flawed, it is generally too complicated to establish. It is even unnecessary since no real threat is present as concerns ballistic missiles. It takes invaluable resources from investments in military measures directed against actual, real threats. Even in the twenty-first century the offensive will always overshadow defence. Second, missile defence is *technologically* impossible: how in any credible way to hit a bullet with a bullet? How to cope with the defensive countermeasures? In the long run, technology will prove that increased speed and precision will add to offensive, not defensive, measures as concerns ballistic missiles. Missile defence is moreover far from being technologically cost-effective. Third, it is *economically* a boundless waste. It is simply too expensive, and does not pay in the long run. It is in other words identical to throwing away money. Fourth, missile defence is negative in terms of *international politics*. It encourages the arms race, it is a detriment to arms control, and it is a threat to the delicate balance of power in world politics and thereby endangers international stability. Its realization could decouple

Europe from the United States. It is generally to be considered as a threat to international peace and security.

This brief and condensed collection of arguments against a ballistic defence system, whether it is called ABM, BMD, NMD, or simply – as is now the normal term – MD, missile defence, includes arguments from the heavy debates in the 1990s and around 2000. The debate in the United States as well as in most other parts of the world has, however, lost momentum and is presently mostly minor and scattered attacks on factual, specific programmes as part of the general missile defence concept. Scholars, politicians and commentators fundamentally disputing the wisdom of developing a MD are relatively rare as compared to the past. 11 September 2001 has resulted in an implicit impact upon the general discourse, despite the fact that this attack had nothing to do with missile defence. It had, however, a lot to do with *defence*. Defence became a key word, materialized in the establishment of the Department of Homeland Security. But also, the notion of defence was enlarged to incorporate extreme offensive measures, e.g. the strategy of preventive attacks.

Strategically senseless?

There are many voices stating that missile defence is strategically senseless. Among the arguments is the claim that missile defence is weakening nuclear deterrence. Although the strategic situation has changed after the Cold War, deterrence should still be an important part of the US military strategy; a missile defence will make deterrence less credible, since it is impossible to establish a missile defence that is 100 per cent effective. The first and fundamental role of the game in a deterrence relationship is to have effective weapons of retaliation.

The counter-argument from the supporters of the official policy is that, first, deterrence is playing a new and less central role in the post-Cold War strategic setting since the central East–West nuclear balance has waned, and second, that any kind of missile defence will strengthen deterrence, as the threats against the United States are expected to come from rogue states supporting terrorism. Without a missile defence – despite this defence not being expected to have 100 per cent protective ability – the United States could be self-deterred not to strike in self-defence, including striking in a preventive or a pre-emptive way against aggressive or expanding rogue states.

Arguments that a missile defence will weaken national as well as international security are extending the general arguments of deterrence further by emphasizing the fact that what counts most in military strategy is the superiority of offence versus defence. In this logic the US has to extend the ability to strike offensively worldwide, be it preventively or pre-emptively. In the twenty-first century the superpower has less need for defence and protection: the open, individualistic society can best be protected by demonstrating the will to remain invincible, and demonstrating the ability to defend itself mainly by offensive means as increased protection may lead to decreasing personal freedom. In contrast to the wisdom of von Clausewitz, who considers defence to be superior to offence due to the fact that defence includes natural factors, e.g. rivers, mountains and oceans, the wars of the new century are increasingly independent of time and space, of territory and geography. To establish missile defence is in this understanding thus not in accordance with the insights of the revolution in military affairs (RMA) the notion that is the leading star of modern American defence, marketed under the term 'transformation'. RMA gives priority to offensive measures. Counter-arguments will emphasize the need for protection against WMD, the biggest threats of all. Protection and defence are relevant and necessary factors in dealing with threats from incoming missiles, factors that indeed make sense in military terms. Again, demonstrating defensive capabilities is necessary for domestic reasons, but is certainly also a military asset in the conflict with the opponents of the new American world order.

Many military-strategic related arguments go like this: missile defence is unreliable, flawed, and has not proved to have any military value. The long history of missile defence has demonstrated a long series of failures, beginning in the 1960s. The only system to become operative was decommissioned a few months after being declared operative. It is still questionable if the system will be effective. It is full of flawed tests, failed basic architecture. Why set up systems that according to any military standards are not living up to even modest expectations? This is considered to be a waste of resources. Even worse, missile defence systems are believed to be a wrong priority. They will swallow up valuable resources that are heavily needed in areas that are of immense importance for the military security of the United States: the war against global terror, the war against – in the typical Rumsfeld formulation – 'the unknown unknowns', the war that has been characterized as World War III.

The counter-arguments are quite simple: yes, the process of establishing a missile defence has been tough, disappointing, and in many ways flawed. Yes, many resources have been used in order to conduct necessary research and development, in order to investigate the possibilities and challenges on the one side and on the other the limits and constraints of the broad field of missile defence. Despite weaknesses, it is, however, worth it. Two factors have to be emphasized. First, two vital experiences and lessons learned as part of the missile defence development process, the closing down of the Sentinel system in 1976 and the SDI announcement in 1983, have demonstrated that it does pay to prioritize critically. Both cases were a manifestation of US strategic and technological superiority.

Second, for domestic reasons any serious attempt to set up protection against WMD carried by ballistic missiles that a growing number of states possess, is worthwhile. It is necessary and unavoidable. And for military reasons the United States, being the dominating military power in the world, has to be in the forefront as regards missile defence. A mutual vulnerability vis-à-vis rogue states has no military meaning. The superpower has to use any military means available and thinkable to reduce vulnerability. To protect itself against terrorists using unknown ways of exploiting American vulnerability and inflicting massive damage on the US homeland and US property is one extremely important priority. But to refrain from attempting to protect against very well-known and possible threats from rogue states or other adversaries, is in US governmental reasoning a policy that cannot be forgiven. Even many analysts who are critical of missile defence are in the situation after 9/11 of being more or less in agreement with this statement. A critical voice such as the director of defence studies at the Cato Institute, Charles V. Pena, is for example emphasizing that:

> Given a strategy of less, rather than more military involvement, and to the extent that a missile defense is technically feasible – proven to be operationally effective (via realistic testing, including against decoys and countermeasures) – a limited land-based ballistic missile defense system designed to protect the US homeland makes sense.[15]

The critical dimension is not directed against the decision of President Bush when he announced the withdrawal from the ABM treaty by arguing that 'Defending the American people is my highest priority as Commander in Chief. Nothing can prevent us from developing effective defenses.'[16]

Technologically senseless?

There are convincing arguments claiming that missile defence is technologically impossible. As already mentioned, the impossibility is demonstrated in the telling metaphor, 'to hit a bullet with a bullet'. On the other hand, technology is developing extremely fast. Tomorrow's technology is today's glint in the eye. Practically everything seems to be possible if one is prioritizing the technological development programme by allocating sufficient resources.

This is certainly true, but there are still barriers that are very difficult, and seemingly impossible, to overcome in the near future. And there is still a very long way to go. It is simply too complicated. Missile defence seems to belong to this group of seemingly unbridgeable gaps in technological innovation and development. The explicit precondition for starting deployment has during all the years been that it was technological feasible and ready for immediate use. This precondition no longer seems necessary. Does the present system function?

To many scientists the present missile defence system is flawed and will not work. A passionate but highly competent critic of the present missile system is MIT professor Ted Postol. He has made a crusade out of exposing the flaws in the system.[17] He and other critics emphasize that the technology needed for an effective missile defence systems still does not exist. All the systems under development are in the early stages of R&D. They will have undergone only rudimentary testing by the time they are deployed, 2004–6. The test conditions will remain far from realistic. Further, none of the X-band radars that are central to the system will be built by 2004.[18] Even if technology worked perfectly, the systems being deployed are vulnerable to countermeasures. These countermeasures are far easier to build than the long-range missile on which they are placed. A comprehensive report from Union of Concerned Scientists and MIT was instrumental in calling attention to this problem. It contributed to President Clinton's decision in 2000 not to deploy the system that the Bush administration is now fielding.[19]

According to Professor Postol and his associates, the claims are that the programme is based on the procedure of first getting the system to work against missiles without realistic countermeasures and then secondly to get it to function against missiles with countermeasures. This situation is similar to plans to build a bridge to the moon. Instead of assessing the feasibility of the full project before moving forward, it is decided to start building the onramps, since that is the part they know how to do.

Another organization, the American Physical Society, has raised serious doubts about the technical effectiveness of the boost-phase weapons under research and development. A 400-page report from a 12-member group under the APS, the largest US association of physicists, released in summer 2003 concludes that a boost-phase anti-ballistic weapon will 'push the limits of what is technically possible'.[20] While the boost-phase weapons might provide some defence against longer-burning liquid-fuelled missiles, they would prove entirely ineffective against faster, solid-fuelled missiles, missiles that North Korea and Iran are likely to possess within the next 10–15 years. The study did not deal with the system to be deployed in 2004 which is intended to target missiles in their midcourse flight. But, as indicated, the Bush missile defence programme is a multilayered programme aiming at incoming missiles in all phases. Characteristic enough for the present discourse on missile defence, the APS study group stopped short of calling the boost-phase part of the missile defence programme 'a waste of money'. The purpose of the group was, according to the co-chairman Professor Daniel Kleppner, 'just to bring the facts forward'. The report had, however, a certain impact on the congressional support for boost-phase systems: the Congress decided to slash the administration's requests in this specific area. It is interesting, however, that another boost-phase programme known as the Airborne Laser, which involves mounting a laser in a Boeing 747 jetliner with the purpose of zapping missiles, received the requested amount of money. This programme, however, is delayed, not least due to weight problems, and testing has been rescheduled for 2005.

The counter-arguments are primarily emphasizing confidence in the technological feasibility of the projects, but the arguments also contain a certain portion of uncertainty. In the Bush administration in general and in the Pentagon in particular there are of course differing opinions of how to prioritize, but all the decision-making units have one thing in common: a defensive missile system *has* to be developed. The United States as the only superpower has no choice in this matter. This way of reasoning is also bipartisan and is, as we have seen, heavily supported by the population. The problem is when and how. It seems to be the common opinion that it is preferable to establish an operative system as soon as possible, even if it is poorly functioning in technical terms, in order to demonstrate – domestically and internationally – that the United States has the will and ability to stand up against a coming threat. A sceptical voice will urge that maybe it is politically and strategically wiser to continue with an innovative R&D programme emphasizing solutions along a broad spectrum of technological possibilities. The simple fact is that the administration has decided to do both.[21]

Therefore, the counter-arguments are all in the same direction: the programme is technically possible because it has to be possible. With enough resources the technical solutions will appear, if not now then in the future. It is necessary to focus on present as well as future threats. To neglect the possibility of countering missile threats is unforgivable. A fast technological fix is, however, not enough. The United States has to be at the technological forefront all the time. And not to deploy operative systems until they are 100 per cent workable and reliable is a failure. One has to admit that systems will never reach the 100 per cent limit. This is politically acceptable and possible owing to the highly positive public discourse on missile defence in the United States.

Economically senseless?

After 9/11, security admittedly has the highest priority. How to get, not 'the biggest bang for the buck', but the most solid security and the most secure protection for the buck is certainly on the political agenda. Should the money be spent on the traditional conventional forces, on the ongoing transformation processes including R&D, on homeland security, on the war on terrorism, on nation building and democracy processes, or rather on missile defence? According to the IISS *Strategic Survey*, the general increase in defence expenditure for the fiscal year 2004 can be estimated as 4 per cent, but the spending for ballistic missile defence has increased by 12 per cent, from $8.1 billion to $9.2 billion. In the future, huge clashes of interest will emerge over, not so much on 'how *much* is enough', but on *how* to spend the money for defence. State and federal interests will clash and there will be clashes between the gap between local and economic interests, between the services, between the bureaucratic organizations and institutions, and between strategic interests on homeland or on global defences. The ongoing wars – the war on terror, the war in Afghanistan and the war in Iraq – will also contribute to growing budgetary problems.

Adding to this there are the ongoing and already mentioned activities concerning missile defence that are denounced as a possible waste of money. We are here referring to the boost-phase efforts that have been heavily criticized by the APS report. Despite the report's negative conclusion, the Pentagon will spend nearly $1 billion in 2004. Also the October 2004 deployment announced by the administration includes 'up to 20 sea-based interceptions employed on existing Aegis ships to intercept ballistic missiles...during the boost and ascent phases of flight'.[22] According to the Center, 'the administration plans to bury its head in the sand and continue to pour billions into a missile defense approach that cannot be effective either now or in the future'.

It is unavoidable that political and administrative controlling pressure will remain on the missile defence projects and programmes. Basic, problematic questions will remain. Is the limited outcome worth the heavy investment? Are the projects well grounded, well researched, well tested, well funded? Is the rush to establish an operational defence system by the end of 2004 immature and built upon a political demand without a proper technological basis? There are several examples of failures and money spent in vain.

How is the administration trying to explain the flawed use of money on missile defence – and how is the current prioritizing justified? Generally the costs for missile defence can be considered a pittance; it is only a small fraction of the defence budget, not to mention of the net federal budget. As a point of departure it has to be underlined that by far the biggest part of the budget for missile defence is reserved for R&D. The budget is $8–9 billion, but less of the one-tenth of this amount is used for what is called MILCON, Military Construction, which is one of two major funding appropriations, the other and totally dominating one being the RDT&E, Research, Development, Test and Evaluation. MILCON deals with the practical work connected with the deployment of systems designed to be operative, primarily the systems to be fielded in Alaska and in California.

First, it is a high priority for the administration not just to focus on the current deployment but also to be highly prepared for the coming years and for the future technological and military position of the United States worldwide. Second, it has to be emphasized that long-term planning for research is extremely difficult. Following a totally prepared schedule will exclude unexpected and valuable technological gains, insights and inventions achieved during the research and development process. The missile project is a unified effort but is certainly not programmed to look for or focus on only a single set of solutions. Third, there may be, or is supposed to be, some dead ends, some flawed projects, and some crucial goals that are not immediately met. But every effort necessary to protect US citizens, including allies and friends, against what is considered to be the most fatal threats, missile attacks with weapons of mass destruction, has to be supported. Missile defence is popular because Americans want to regain as much as possible their sense of invulnerability and are willing to bear the costs for doing so.

Internationally senseless?

As has been indicated, one of the most crucial and persistent arguments against the deployment of missile defence has been the negative effects upon the development of world politics and on international stability. It was the general attitude among the nations that although the arms control agreements from the Cold War had lost some of their original political importance and focus, they were still relevant and worth maintaining. The open signals from the United States towards withdrawing from the ABM treaty intensified the general opposition to the US missile defence plans. One can even speak of an international 'united front' in this respect. The arguments were that missile defence would weaken deterrence, weaken international stability and security, ignite an arms race and create fear of US dominance. So the united front encompassed most European states, including Germany, France, Russia, China, India, and practically most other states in the world. The United States was thus pretty alone in this case.

We have already dwelt upon the arguments concerning the international realm for and against missile defence. The United States fought almost alone, defending its right to defend itself against incoming ballistic missiles that could or probably would contain weapons of mass destruction, be it unauthorized, rogue state delivered, or coming from elsewhere. The US arguments claimed that a missile defence (referring mostly to the US national missile defence concept) would strengthen deterrence, which was still an important part of US strategy; further, missile defence would increase strategic stability and international peace and security; it would dampen the arms race; and no US dominance was intended, but as the United States position in the international system as a fact of life was dominating owing to its capabilities, it was its national as well as international duty to protect itself from attacks or from blackmailing due to the credibility of possible threats of offensive actions with ballistic missiles.

The whole international setting changed, however, as a result of 9/11. The structure of the international system remained unchanged, but fundamental foreign and security policy transformations took place. First, the United States effectively, declaration-wise, widened the concept of missile defence from the narrow national missile defence into a global missile defence programme intended to cover allies and friends, and focusing not just upon ICBMs, but also attempting to include short-range ballistic missiles, medium-range ballistic missiles and other kinds of missiles; also protection against cruise missiles should be included. Second, as part of the general international acceptance of the war against terrorism under US leadership supported by the United Nations, an acceptance already expressed by the generally approved declaration the day after 9/11 in the French newspaper *Le Monde:* 'We are all Americans', an international discourse became dominant. The shorthand version was human civilization versus terrorists. Very quickly, however, contrasting interpretations of how to interpret or define terrorists appeared. Russia and China were both quick to exploit the window of opportunity suddenly opening up: the possibility of combining the establishment of structurally necessary close relations to the superpower with a closer international recognition of the way in which the two countries behaved vis-à-vis what they considered *their* terrorists, i.e. in the case of Russia the uprising and fighting in Chechnya, and for China the ongoing fight against the Muslim uprising in the Xinjiang province.

The result was that Russia reacted in a very low-key way to the American decision in December 2001 to declare that they intended to withdraw from the restraints of the ABM treaty, a withdrawal which according to the treaty's Article 15 could enter into force after six months, i.e. in June 2002. Likewise China – despite protests and modest threats of countermeasures – tacitly accepted the new state of play. In Europe the situation developed along the same path. In summer 2001 President Bush visited Europe in order to gain support for the administration's missile defence programme, resulting in rather reluctant and hesitating if not simply negative attitudes on the part of the European allies. This changed fundamentally after 9/11. At the Prague NATO summit in 2002 it was decided to launch a new study to develop strategic missile defence systems able to protect the whole of the European territory.[23] The United States had expressed its interest in assisting the European NATO partners to deploy missile defence systems that were intended to be part of a truly global system of defences,[24] based on technical insight into the American technologies and possible European industrial participation. It is, however, doubtful if the European governments, most of whom are in the process of fundamental transformations of their military forces that probably in only a very few cases will imply an increase in military expenditure – rather the opposite – will choose to give priority to missile defence efforts. But European states are, nevertheless, directly related to the US missile defence plans.

The US administration formally requested the British and Danish governments to permit the upgrading of radars at Fylingdales base and Thule base, Greenland, respectively.[25] This happened on the same day, 17 December 2002, as the administration announced the decision to begin deployment of the missile defence system in 2004. In February 2003 the

UK approved the upgrading and Denmark followed suit in May 2004 after a period of a public debate that never really materialized. Japan as well as Australia has indicated a willingness to join the US in the attempt to establish regional missile defence systems. As already mentioned, China and Russia are in different ways trying to adapt to the current situation with the United States as the worldwide dominating anti-ballistic missile power.

Explaining the US missile defence programme

So why is it that missile defence has gained the highest priority in the development of US defence capabilities? The basic claim is that missile defence is closely connected to the objective of sustaining the US superiority in the international system.

Theory in the neorealistic tradition will tell us that the United States – like other states – will strive for survival. There is a reason for having an effective defence reflecting the US relative capabilities worldwide. Also theory tells us that the United States is aiming at keeping its current relative position in the international system, i.e. its position as the only superpower. It is as a part of these simple assertions that one has to find the explanation for the development of a US missile defence. The basic claim is that the main agenda for the United States is not the specific missile defence programme. The programme is indeed important, but it has to be assessed in a much broader context. For the United States it is – in the view of the international and global challenges of the twenty-first century – a priority to maintain and widen the general technological and military superiority worldwide and thereby its relative political position.

Missile defence has to be assessed, first, as part of the general strategy of the United States, second, as part of the specific military-strategic-technological development, referred to as RMA, the revolution in military affairs, and third and finally, as part of two crucial interrelated dimensions: the dimension of cyberspace and the dimension of space, the mantra being 'He who commands the cyberspace as well as the space commands the world'. Expressed in a simple way: in the strategic setting missile defence refers to the notion of *superiority*, in the technological setting missile defence refers to *innovation*, and finally in the space and cyberspace setting it refers to *exploration*.

First, an important part of the National Security Strategy (NSS) of September 2002 is – besides the strategy of prevention and pre-emption – the *dissuasion* strategy, a strategy that implies a continuing worldwide dominating military superiority. In part 11 of NSS it is stated, 'Our forces will be strong enough to dissuade potential adversaries from pursuing a military build-up in hopes of surpassing, or equalling, the power of the United States'.[26] This part of the strategy implies that the United States will never accept a kind of return to bipolarity, which reigned during the Cold

War, including the acceptance of military parity as a basis for the strategic relationship. Neither will the United States consent to a multipolar structure, implying that a combined force of US adversaries could equal or even surpass the United States in military power. Having the dissuasion strategy in mind as a cornerstone of the national security strategy, there will be an implicit requirement to demonstrate that dissuasion has a concrete substance. The means of dissuasion is to keep on the alert all the time by having the most sophisticated, advanced programme of military research and development in order not to be taken by surprise by any potential adversary. Or, as stated in the strategy paper, 'to be menaced by catastrophic technologies in the hands of the embittered few'.[27] The US can never stop being at the forefront of any technological development that can be used for military purposes. This is a heavy burden for the United States, as it can never give maximum priority to solving a current threat. There always have to be continuing efforts to look behind the actual situation, taking into consideration the coming medium-term, and certainly also the long-term military-technological development. In this context, missile defence is an important factor: the United States is – according to the dissuasion strategy – aiming at a long-term programme that is extremely all-embracing and that cannot exclude any feasible technological solution to the missile defence problem.

Second, all statements concerning the American military development begin and end with reflections and invocations dealing with the concept of transformation. Transformation is the official term for RMA, the revolution in military affairs. It involves a change of culture, of thinking as regards war and peace, roles, tasks and performances of the armed forces, requiring new strategies, tactics, organization, equipment, education. Transformation is characterized by military operations conducted with knowledge, speed, precision, lethality and surprise. It involves a reassessment of the types, locations, numbers and capabilities of the US military forces worldwide. The forces have to be more agile, lighter, and flexible. The Chairman of the Joint Chiefs of Staff, General Richard B. Myers, is referring to a whole spectrum of changes, pointing to an acronym, DOTMLPF, or doctrine, organization, training, material, leadership, personnel and facilities. Normally one is focusing on the M-factor – the material. But more important than technology is, according to Myers, the intellectual, the mental component in the transformation.[28]

The crucial problem is that transformation or RMA is an ongoing process, where the US military 'still is grappling with the implications of this "megatopic"'.[29] The transformation is 'still quite incomplete'. The enormous complication is that the US military posture in many ways is still a legacy of the Cold War.

What are the relations to missile defence programmes? Looking at the mission for the missile defence, namely, 'to develop and field an integrated BMDS, Ballistic Missile Defense System capable of providing a layered defense for the homeland, deployed forces, friends, and allies against ballistic missiles of all ranges in all phases of flight',[30] it seems to be defined rather narrowly and precisely. In theory the ways and the means to achieve the ends are open. Looking at the appropriations for the R&D for missile defense for the coming years, there seems to be some sort of coherence inside what insistently is referred to as an integrated project. This integration notion is, however, in opposition to the general philosophy, also emphasized by the President, namely, 'to establish some kind of workable defence, to have a start, some point of departure, but allowing for development dealing with the "unknown unknowns"'. In order to catch the spirit of transformation, any kind of incremental change is considered negative. Taking existing systems and, in small jumps, making them better is not transformation.[31] And transformation is particularly relevant as regards missile defence. This requires a couple of important factors: first of all innovation. Innovation is the mantra for practically all policy areas of the United States. An example is the claim that innovation is the key element of US foreign policy. As it is expressed, 'The policy must respond, through innovation, to changes in the environment of both threat and opportunity'.[32]

Innovation is crucial for missile defence since it is fundamentally based on an attempt to set up a kind of defence that until now has never worked. Innovation leading to the invention of completely new weapons systems that can change the face of war is basically what transformation is about. The main examples from the last 50 years will obviously include the A- and the H-bomb, the nuclear-tipped intercontinental ballistic missiles, and the Polaris submarine-launched missiles, the SLBMs. These systems fundamentally changed the face of war. In a strange way also the SDI project had a similar effect: it certainly changed the face of the so-called virtual war between the superpowers resulting in the Soviet voluntary surrender. The missile defence system is also expected to change the face of war, not least based on the invention of brand-new technologies, for example the hit-to-kill concept, using kinetic energy as a weapon, or the exploitation of directed energy, primarily lasers, used for fighting purposes but also for communications. A further attempt to change the face of war lies in the planned first part of the layered missile defence, the boost phase. The technical innovations that are necessary to hit a missile in its extremely vulnerable, but also extremely short, initial phase will in a way transform the general notion of the term defence as opposed to offence. Exceptionally fast and comprehensive access to relevant information and data is a precondition for exercising a defensive counter-attack; further, anti-missile weapon platforms have to be rather close to the enemy missile launching position, which again will require large amounts of such weapon platforms that have to be quickly deployable.

These considerations are in full accordance with the general transformation thinking in the United States. It deals with the question of the presence of US forces. Presence indicates that military capabilities in certain locations are needed to provide either reassurance to allies and partners or to deter and/or dissuade those who would challenge the United States, its allies or its interests. Another aspect is the pre-positioning of equipment and supplies. Pre-positioned equipment should facilitate movement of US forces to areas where they are needed.[33] With these technologies made operative, the United States makes available to itself a great variety of strategic choices. Prevention and pre-emption are already important parts of the options: the possibility of a strike against a missile in the boost phase is very close to – but still very different from – a launch on warning act, which again could be perceived as a kind of defensive measure.

Also, transformation has to do with the hidden and the unknown threats. As regards missile defence, the types, the characteristics and the performance of the different missiles and the countermeasures against missile defence are well known to the US military. Only few surprises are expected. Still, the missile defence research and development is aiming at collecting and eventually exploiting the broadest knowledge possible, not primarily regarding the possible current or future threats, but certainly regarding how most effectively to meet and counter these threats.

Third, the two interrelated 'new' dimensions of warfare, space and cyberspace, have to be mentioned, contributing to explain the crucial position of missile defence. Space has in the past been a priority for US defence, but has experienced a kind of neglect in the last couple of years. Vice Admiral Cebrowsky, responsible for force transformation, in December 2003 characterized the situation in the following way, emphasizing 'the need for a shift from *supplier-centered intelligence* to *demand-centered intelligence*. In this area is the whole idea of space.' The Department of Defense must be more adaptable in space. He noted that there were 38 launches of micro-satellites in the last few years and that the United States did not participate in any of them.[34] A shift in the priorities is certainly under way. Secretary Rumsfeld stated a few days later that 'the importance of space and missile defense cannot be overstated'. Further he emphasized that 'the importance of space will only increase in the future. Space is fundamental to modern warfare and vital to US interests.'[35] Cyberspace is the closely related factor: space is the necessary precondition for the global network, cyberspace. And as is known to everybody working with computers, cybernetworks are extremely vulnerable to attacks. Therefore the United States is using a fair amount of resources to secure not only the military cyberspace but also the vital civil networks. The coming years will see an expanding effort to explore and thereby to be able better to defend cyberspace and space. An obvious aim for the US is to be dominant in space as well as in cyberspace. Major General Blairsdell of the Air Force Operations stated in

March 2003: 'We are so dominant in space that I would pity a country that would come up against us.'[36]

Space is a tricky concept. Space can be defined as everything above the atmosphere, normally defined as reaching up to 100–110 km from the surface of the Earth. Space includes satellites, space vehicles, and ballistic weapons that are returning to the Earth's atmosphere after having been flying outside in their midcourse phase.

Outer space is demilitarized as regards weapons of mass destruction, meaning that it is forbidden to deploy such weapons according to the Outer Space Treaty of 1967. There are no intentions from the United States to violate this international treaty. Nuclear weapons and other weapons of mass destruction have no relevant role to play in space. Nuclear weapons would have devastating effects if brought to explosion in space as well as in outer space, stopping all electronic signals and equipment. This could emerge as a new threat. In autumn 2003 President Bush issued a plan to revitalize the peaceful exploration of outer space, preparing a mission to Mars to take place in 10–15 years. Although there is no direct connection between the civil efforts to investigate outer space and the military programmes that are restricted to space, both endeavours are part of a broader concept of staying in and enhancing the position of being at the forefront as regards general technological R&D.[37]

So, space and as a consequence cyberspace will, however, increasingly be the main battlefield and place of competition between the major powers, following the assertion that in the twenty-first century, 'He who commands the space will command the cyberspace. And he who commands the cyberspace will command the world.' In this competitive realm the United States has – and intends to keep – military and technological superiority. This is done through massive and diversified R&D efforts in military and military-related technology now and in the future.

To sum up, although President George W. Bush on several occasions has stated that 'America's development of a missile defense is a search for security, not a search for advantage',[38] we claim that a superior explanation of the current US missile defence programme taking its point of departure in a neorealistic theoretical understanding, will state that the decisive factor is maintaining the American strategical-technological *superiority*. Evidence is found the US National Security Strategy of 2002, accentuating dissuasion strategy. RMA is supporting missile defence in terms of *innovation*. Space and cyberspace issues support by emphasizing the *exploration* factor. All these objectives are enshrined in the dissuasion strategy.

Conclusion

The objective of missile defence in a narrow sense is to reduce American vulnerability against WMD attacks. One could add that to achieve this objective for around $10 billion a year, a relatively small fraction of the American defence budget, is impressive. The project is, however, expanding. The actual deployment and the current R&D projects are only the beginning. The need for new technologies and the demand for continuously staying in the frontline of military-technological research and development will for the United States imply still expanding programmes dealing with missile defence in a broad understanding. The main argument of this chapter is that besides the need for meeting current new threats and challenges, there are long-term agendas behind the US missile defence programme: for the United States to ensure a continuing and increasing strategic-political-technological worldwide superiority, while at the same time being able to assist allies against WMD threats from rogue states and units.

To explain in full the efforts to establish American missile defence programmes, one has – owing to their visible shortcomings and widespread scepticism from different constituencies – to go beyond the immediate defensive needs for the United States to defend against rogue states, to reassure against unintended firing of missiles, to deter rogue states from developing intercontinental missiles, to increase deterrence in a crisis, and to protect US and allied forces against missiles. The shortcomings and scepticism refer to missile defence weaknesses on the military, technological, economic and international dimensions. Since its beginning after the introduction of the ICBM, missile defence has been closely attached to superior political-strategic considerations and rationales. The current missile defence programme, developed in a unipolar world, is thus – according to theory – to be explained in terms of survival, relative power, and the maintenance of the relative position in the international system. Missile defence has to do with an immediate search for security, but it is primarily a long-term weapons systems programme aiming not least at maintaining the strategic-military superiority worldwide. The main statements in the 2002 National Security Strategy empirically support the long-term objective, not least by what can be referred to as the dissuasion strategy. A marked expression and manifestation of this part of the strategy is growing emphasis upon keeping technological superiority, not least by RMA-based innovation and space and cyberspace-based exploration.

This analysis has demonstrated that the US missile defence programme, despite military and strategic deficiencies, despite technological flaws, despite accusations of waste of money used for seemingly far-fetched in the long run and non-workable technologies, and despite international resistance, will continue to have a high political priority. Further, based on theoretical assertions and empirical evidence from the primary political declarations, the analysis suggests that missile defence has been given the highest political and economic priority, not primarily to provide the United States and its allies with a defensive, protective shield, but primarily to have missile defence as an important political vehicle as part of the far broader and long-term project: by all means to maintain the United States as the unrivalled and matchless superpower, unsurpassed technologically and militarily.

Notes

1 On 10 December 2003 the US Secretary of Defense, Donald Rumsfeld, declared that defending the United States and its allies from ballistic missiles laden with weapons of mass destruction 'is now America's highest priority'. The Secretary of Defense stated that 'we had entered a new age that may well be the most dangerous America and the democracies of the world have ever seen'. Already in 2002 President George W. Bush had announced a similar invocation. On 17 December 2002 he stated, 'The deployment of missile defenses is an essential element of our broader efforts to transform our defense and deterrence policies and capabilities to meet the new threats we face. Defending the American people against these new threats is my highest priority as Commander-in-Chief, and the highest priority of my administration.'

2 Department of Defense News Briefing, Secretary of Defense Donald Rumsfeld, 26 August 2004.

3 The International Institute for Strategic Studies, *Strategic Survey 2002–03*, London: IISS, 2003, p. 30.

4 Ibid., p. 29.

5 See Missile Defense Agency, *Ballistic Missile Defense System*. Online. Available HTTP ⟨http://www.defenselink.mil⟩.

6 In February 2002 the Missile Defense Agency announced three phases, C1, C2 and C3. The C1 system should by 2005 be able to defend against a handful of missiles coming from North Korea. The C2 system will in 2007–10 be able to meet ballistic missile threats coming from Iran. The C3 system is supposed to annihilate a Russian ICBM force of less than 1,200 nuclear warheads. Ingemar Dörfer, *Ballistic Missile Defence. Det amerikanska programmet: Säkerhetspolitiska konsekvenser*, Stockholm: FOI, 2002, p. 20.

7 The first phase, the boost phase, is the phase where it is gaining the velocity needed to reach the target. This phase lasts between one and five minutes, depending on the range of the missile. The end of the phase is when the missile is exiting the Earth's atmosphere or, in the case of shorter-range missiles, reaching the fringes of outer space. The best solution for attacking a missile is in the boost phase. The missile is extremely vulnerable in this phase as it struggles against Earth's gravity. It is still attached to the warhead, so possible decoys are not a problem, and not least, if the missile is carrying chemical, biological or nuclear weapons, the debris will most likely fall on the country that launched the missile. At least, it will certainly not have obtained enough velocity to reach its intended target. Because of this, it is not critical, according to the Missile Defense Agency, to completely destroy the warhead of the missile. The defenders' problems are, however, huge. First, the boost phase is relatively short. Second, an interceptor missile has to be very close or extremely fast to catch up with the accelerating missile. The capability of intercepting a missile in the boost phase is important as it destroys the missile regardless of range and intended target. Boost-phase defense can provide a global coverage.

The second phase is the midcourse. It is the longest part of the flight of the missile, as long as 20 minutes in the case of ICBMs. During this phase the missile is coasting, or freefalling towards its target. This phase allows the largest opportunity to intercept an incoming missile. The missile now follows a predictable path. This implies that several interceptors could be launched with a delay between them to see if the first ones were successful. Also this means that fewer interceptor sites are needed to defend larger areas, since the interceptor has a longer time to engage. The problems with this phase are the countermeasures against a defensive system that the attacking missile can deploy. A midcourse defence is aimed at being able to provide missile defence for a region or regions.

The third phase, the terminal phase, is the phase where the missile's warhead re-enters the Earth's atmosphere. This phase lasts less than a minute for ICBMs. The defensive systems have to be very close to the missile's target in order to intercept it. Countermeasures play a minor role in this phase. Decoys usually fall more slowly than the warhead and are burned up as they re-enter the atmosphere. Defensive systems designed for this final phase are most effective in protecting troop concentrations, ports, airfields and missile staging areas. Terminal defences are thus mainly designed for protecting localized areas. Missile Defense Agency, *MDA-facts, MDA, In-depth. Making Ballistic Defence a Reality*. July 2003. Online. Available HTTP:⟨http://www.acq.osd.mil/bmdo/bmdolink/html⟩.

8 The development of the American missile defence was clearly driven by two factors, an internal action–reaction process, indicating that development of defensive missiles was a reaction to one's own offensive weapons, and an external action–reaction process, indicating that the US MD was a reaction to Soviet weapons development. The US reacted to the Soviet missile defence systems, deployed around 1963, and also to the rapid development of Russian ICBMs in the 1960s. Many resources were used in the US to enhance the penetration ability of the long-range missiles. The United States developed advanced weapons substituting the SLBM Polaris and the ICBM Minuteman I and II. From 1970 the new Poseidon and the new Minuteman III were introduced. Both were equipped with penetration aids and were MIRVed, i.e. they had 3–10 independently guided warheads. These systems were considered successful.

9 John H. Herz, *International Politics in the Atomic Age*, New York: Columbia University Press, 1959.

10 See Raymond L. Garthoff, *Detente and Confrontation: American–Soviet Relations from Nixon to Reagan*, revised edition, Washington, DC: Brookings Institution, 1994, pp. 325–359.

11 After the end of the Cold War, American missile defence continued. Secretary of Defense in the Clinton administration Les Aspin in May 1993 changed the name of the Strategic Defense Initiative Organization to Ballistic Defense Organization (BMDO). On this occasion the Secretary stated that the change of name marked the end of the SDI decade and that SDI had played a decisive role in ending the Cold War. In fact it is possible, according to a well-known scholar, to put a date to the end of the Cold War: 21 September 1989. On this date the Soviet Foreign Minister personally delivered a letter to President Bush from President Gorbachev, containing a couple of important Soviet concessions on the Ballistic Missile Defence issue regarding the procedure of retreating from the treaty, demolishing the contested Krasnoyarsk radar, regarding verification, also including concessions in the negotiations on START, the Strategic Arms Reduction Talks. The two parties were still not in agreement on the ABM issue. But the letter was an indication of throwing in the towel. The days of the Soviet Union as a superpower were numbered.

12 The speech was a surprise. Only very few people knew about its content. A popular interpretation says only three: besides the President, Nancy Reagan and Edward Teller, the father of the H-bomb. Obviously a group of military, strategic and technological experts had discussed and drafted the idea. The vast bureaucratic and political establishment, however, had not been involved.
13 Dörfer, op. cit., p. 3.
14 The International Institute for Strategic Studies, op. cit. p. 28.
15 Cato Institute, *Missile Defence: Defending America or Building Empire*. Online. Available HTTP: ⟨http://www.cato.org/dailys/07-13-03.html⟩.
16 Ibid.
17 Union of Concerned Scientists, *Postol vs. the Pentagon*, 23 June 2003. Online. Available HTTP: ⟨http://www.nuclearfree.lynx.co.nz.diaopposed.htm⟩, pp. 1–3.
18 Ibid., p. 1.
19 Ibid.
20 Bradley Graham, 'Questions on Missile Defense Plans', *Washington Post*, 16 July 2003, p. A02. Online. Available HTTP: ⟨http://www.washingtonpost.com⟩.
21 Therefore the focus has all the time to be on the *process* rather than on the immediate *outcome*. As was emphasized by the Missile Defense Agency in July 2003: 'The Department of Defense established a single program to develop an integrated Ballistic Defense System (BMDS) under the Missile Defense Agency. And while there is only a single BMDS, there is no final or fixed missile defense architecture. We are employing a spiral development approach to incorporate upgrades to the BMDS, the goal of which is to 1. Field an initial capability in 2004–05, 2. Add networked, forward deployed ground-sea, and space-based sensors to make the interceptors more effective in 2006–07, 3. Add interceptors, 4. Add layers of increasingly capable weapons and sensors, made possible by emerging technologies'. Online. Available HTTP: ⟨www.acq.osd.mil/bmdolink⟩.
22 Matt Martin, *Laws of Physics vs. Boost-Phase Missile Defence, Round 1*, Center for Arms Control and Non-Proliferation, 23 June 2003. Online. Available HTTP: ⟨http://www.armscontrolcenter.org/archives/000236.php⟩ (accessed 14 October 2004).
23 The International Institute for Strategic Studies, op. cit., p. 39; here it is stated that alliance officials leapfrogged over tentative plans to study theatre missile defences to launch the study.
24 Ibid.
25 Ibid.
26 National Security Stategy of the United States of America, 17 September 2002, p. 30. Online. Available HTTP: ⟨http://www.whitehouse.gov/nsc/nss.html⟩ (accessed 14 October 2004).
27 Ibid., ch. 1, p. 1.
28 Jim Garamone, 'Transformation Part of War on Terror, Myers Says', *American Forces Press Service*, 3 December 2003. (a) Online. Available HTTP: ⟨http://www.dod.mil/news/Dec2003/n12032003_200312036.html⟩ (accessed 14 October 2004).
29 Jim Garamone, 'Global Military Posture's Part in Transformation', *American Forces Press Service*, 2 December 2003. (b) Online. Available HTTP: ⟨http://www.defenselink.mil/news/Dec2003/n12022003_200312022.html⟩ (accessed 14 October 2004).
30 Statement from Missile Defense Agency, 11 December 2003. Online. Available HTTP: ⟨http://www.acq.osd.mil/mda/mdalink/html/mdalink.html⟩.
31 Garamone (a), op. cit.

32 Dr. Mitchell B. Reiss, *Innovation and Continuity in US Foreign Policy*, 6 November 2003. Online. Available HTTP: ⟨http://www.state.gov/s/p/rem/2003/26334.htm⟩ (accessed 14 October 2004).
33 Garamone (b), op. cit.
34 Jim Garamone, 'Looking for Technology, Doctrine to Transform the Face of War', *American Forces Press Service*, 3 December 2003. Online. Available HTTP: ⟨http://www.dod.mil/news/Dec2003/n12032003_200312033.html⟩ (accessed 14 October 2004).
35 Gerry J. Gilmore, 'Space, Missile Defense Essential to Defence, Rumsfeld Says' *American Forces Press Service*, 10 December 2003. Online. Available HTTP: ⟨http://www.defenselink.mil/news/Dec2003/n12102003_200312108.html⟩ (accessed 14 October 2004).
36 Rudi Williams, 'US Dominance in Space Makes General "Pity the Enemy" ', *American Forces Press Service*, 12 March 2003. Online. Available HTTP: ⟨http://www.defenselink.mil/news/Mar2003/n03122003_200303127.html⟩ (accessed 14 October 2004).
37 As regards space, it is, according to STRATCOM commander Ellis, 'essential to everything we do'. In 2002 the Defense Department built a new unified command by combining the missions and strengths of US Space Command and the former US Strategic Command. The new STRATCOM, one of the nine US unified commands, is headquartered in Nebraska. It is the command and control centre for US strategic forces and it controls military space operations, strategic warning and intelligence assessments. It provides operational space support, integrated missile defence, global C4ISR and specialized planning expertise to the joint warfighter. In the opinion of the the commander, 'Only by integration of the commands's aggregate strengths will STRATCOM bring its entire range of global capabilities – space, missile defense, planning, communications, information operations, kinetic and nonkinetic strike, and intelligence – and ensure the US military stays one step ahead of our adversary.' Major John Paradis, 'Strategic Command Missions Rely on Space', *American Forces Press Service*, 26 September 2003. Online. Available HTTP: ⟨http://www.defencelink.mil/news/Sep2003/n09292003_200309297.html⟩ (accessed 14 October 2004).
38 Remarks by Secretary of Defense Donald Rumsfeld, Fort Lesley J. McNair, Washington, DC, 1 May 2001. Online. Available HTTP: ⟨http://www.defenselink.mil/speeches/2001/s20010501-secdef.html⟩ (accessed 14 October 2004).

4

ROADS NOT (YET) TAKEN

Russian approaches to cooperation in missile defence

Alla Kassianova

Introduction

As a nuclear power, Russia at all times had to take account of the proper mix and level of its offensive and defensive capability. Up to 2001, it chose to do so within the legal framework of international agreements of which the ABM treaty was the centrepiece. Embrace of the full-scale Ballistic Missile Defence (BMD) programme by the US, abrogation of the ABM treaty, and the introduction of the pre-emptive doctrine into the US military and security policy all compel Russia to think about the role of missile defence in its own security strategy. As the United States is inviting its 'allies and friends' to join the BMD project, Russia faces a dilemma in determining the possibility of its own cooperation with the US in this area.

On the one hand, nuclear deterrence continues to lie at the heart of the Russian military doctrine and security strategy. The mentality of confrontation rooted in the strategic culture of the past Cold War is slow to ease its grip over the minds of the Russian political and military leadership. This mentality favours a nuclear deterrent whose independence is a prerequisite of its reliability. According to this logic, the US role can be fathomed at any point in the range between 'implicit referent/counter partner/opponent'. With missile defence an integral part of the overall military capability, cooperation with the former strategic enemy is at best difficult for either side. On the other hand, the Russian leadership is keenly aware of the national technological downturn and the general economic weakness. It realizes that cooperative relations with the US in missile defence could bring benefits in the strategic, political and technological fields. For many years, the objective of 'strategic partnership' between Russia and the US has been given a rhetorical commitment. Yet the actual position developed by Russia in relation to the US BMD programme could not be easily characterized as cooperative.

This chapter seeks to map out and explain the contradictory nature of the Russian missile defence policy. The analysis is based on the assumption that Moscow (1) is considering options of further developing the existing system of missile defence, and (2) admits the possibility of entering into cooperation with other countries or organizations. The chapter singles out three potential (collective) partners most seriously considered by the Russian decision-makers in the context of cooperative schemes of missile defence development: the Commonwealth of Independent States (CIS) members of the Collective Security Treaty Organization; Europe (the European Union/NATO); and the US. Of these three, the analysis is focused on the United States as the principal and in fact unavoidable actor in international BMD politics.

The analysis starts with a description of the material assets in the field of strategic missile defence inherited by the Russian Federation from the Soviet Union and upgraded during the late 1990s and early 2000s. The chapter documents the visible growth of the state's attention to the strengthening of the air and missile defence capabilities. It goes on to examine the relations established between Russia and prospective partners in the development of MD, presenting each option in the light of achievements and failures/problems of the respective relationships. In the course of this, it registers a progressive overlapping of air, theatre and strategic missile defence capabilities, further complicating military and political decisions in this field.

In addressing the question of the viability of the prospect for Russian cooperation with the US, the next part proceeds in two steps. In the first step, considerations of hard security, technological strength and status are presented as main areas of concern that provide rationales bearing on the decision-making. Then the discussion identifies groups interested in the issue of missile defence (the military, politicians, and the defence industry) and weighs their attitude to the prospect of cooperation with the US as well as their relative influence on the decision-making process. The analysis concludes that in Russia, such important domestic factors as a sustainable pro-Western political tradition, pragmatic public opinion, technological development incentives, and interests within defence-industrial circles can still be configured into a 'win set' for a cooperative approach in missile defence. However, the domestic layout hinges on the realities of the international environment, many of which, especially in the military and strategic sphere, continue to feed the confrontation-era mental schemes still influential in Russian policy-making.

Soviet legacy reactivated: Russian strategic and other missile defence capabilities

The Moscow ABM system

Soviet nuclear legacy makes Russia the only power whose strategic potential is in any way commeasurable with that of the US. In fact, throughout the last three decades Russia had been the only power with an operational system of strategic MD in the world. In the USSR, missile defence had a long tradition. In March 2001 the Russian defence-industrial community commemorated the fortieth anniversary of the first ever successful intercept of a ballistic missile. Back on 4 March 1961, the missile was coming in from 2,000 km away at a speed of over 3 km per second and had its warhead fully destroyed by the V-1000 interceptor with an ordinary explosive at a height of 25 km over the Sary Shagan testing rage. Following the success of the experimental phase, in the early 1960s the Soviet government set the task of developing an operational system of MD that was developed and eventually deployed in consecutive modifications to protect the Soviet capital.[1] The Anti-Ballistic Missile (ABM) treaty was signed in 1972 when the first incarnation of the Moscow ABM system was already under test. The characteristics of the system were undergoing continuous modernization in response to the sophistication of the US offensive missiles. The latest generation of the system, designated A-135, was formally completed in 1989 but put on combat duty much later in February 1995.[2]

The system of radar support for the Moscow ABM was continuously evolving starting from the late 1960s when the earliest battle management radar Dunay-3 ('Dog House') in Kubinka entered operational use, followed in the 1970s by a second, Dunay-3U ('Cat House'). By 1978, both Moscow ABM radars were upgraded to be integrated with a larger early warning network that had been operational since 1970. Its two Dnestr-M ('Hen House') radars in Olenegorsk, Kola Peninsula, and Skrunda, Latvia, with a command centre near Moscow, watched the northwestern direction to provide the early warning information to the Moscow ABM.[3] Later on the Moscow ABM system was enhanced by the multifunctional phased-array Don-2N ('Pill Box') radar that was under construction for about a decade and reached full operational capability around 1989. Designated as a battle-management radar, it has also been used in an early warning and space surveillance capacity. Finally, the operation of A-135 relies on a command, control and computing centre with high-power computing systems developed in the mid-1980s and later upgraded in the early 1990s and 2000s.[4]

The military utility of the Moscow ABM has always been a contested issue. Even at the level of design, its capacity is limited to non-massive nuclear attack by unsophisticated missiles. Consecutive modifications increased the survivability of the components of the system (concealing the interceptors into the hardened silos) and the reliability of the intercept (improving the radar support, elaborating a two-layer architecture and the performance of the missiles). In the US military and intelligence community its effectiveness has generated a range of dissenting opinions – from blunt dismissals to cautious reservations.[5] The Russian sources range from reports that the system is operational and alert to private voices in favour of dismantling it, as it makes no sense to keep an obsolete and expensive system.[6] As the interceptors were designed to carry nuclear charges, additional concerns arise as to the effect of the intercept explosion on the people, buildings and complex equipment on the ground.[7] In the late 1990s there were a number of reports based on cursory official statements that warheads with ordinary explosives had replaced the nuclear warheads on the Moscow ABM interceptors.[8] However, the information was very scarce and perhaps deliberately left room for speculation. It can be stated beyond a doubt that preserving the deployed Moscow ABM system serves political as much as military purposes.

In the late 1990s the A-135 system around Moscow was supplemented by S-300PMU2 (*Favorit*, or SA-10 'Grumble') air defence systems with an anti-tactical ballistic missile (ATBM) capability. The *Favorit* air defence system with a range of 200 km belongs to the well-known S-300 series deployed in Russia, some of the CIS countries, and in limited numbers abroad.[9] The more advanced S-400 (Triumph) system, after several years' delay, in mid-2003 successfully passed its state tests but its serial production has been put off because of lack of funding.[10] Specifically, the funds were lacking to test the new S-400 missile designed with an effective range of up to 400 km and capable of hitting new-generation air-to-surface missiles and AWACS aircraft. The existing Triumphs carry the older missile of the S-300 series. In 2003, Triumph was approved for serial production and its deployment to Russian troops was scheduled to start in 2004.[11]

The early warning network: ground-based

The Soviet Union invested much effort into building an early warning system combining ground and space components (*Sistema Preduprezhdeniya o Raketnom Napadenii*, SPRN). During the late 1960s and the 1970s, several radars of the Dnestr and Dnepr types, designed for detection of attacking ballistic missiles and space surveillance, entered service in geographical locations to cover the more probable directions of attack: in Olenegorsk (Murmansk Oblast), Skrunda (near Riga, Latvia), Mishelevka (Irkutsk Oblast), Balkhash (Kazakhstan), Mukachevo and Sebastopol (Ukraine). In 1972 the developers proposed a concept of an integrated system that used unified information and data-processing principles and envisioned linking the ground and space tracking assets into a seamless network.[12] By the late 1970s, the ground radar network built on the periphery of the Soviet borders provided nearly complete missile warning coverage of the attack-threatening directions.[13] In the mid-1980s, two new-generation radars of the Daryal ('Pechora') type with expanded power and detection range were made operational in Pechora (northern Russia) and Gabala (Azerbaijan). An ambitious plan maximizing the potential of the earlier and newer generation stations was developed and construction of new Daryal-U and Daryal UM (modernized) radars in Yeniseysk (Krasnoyarsk), Balkhash (Kazakhstan), Mishelevka (Irkutsk), Skrunda (Latvia) and Mukachevo (Ukraine) was started. The meter band wavelength Daryals were to be complemented by the decimetre band Volga radars to provide for dual-band coverage along the entire periphery of the Soviet Union. In the mid-1980s work began on building the Volga radar near Baranovichi (Belarus) with an early warning capability against launches of intermediate-range ballistic missiles based in Europe.[14]

From the mid-1980s, the progressive development of this system slowed down owing to the general economic decline, then halted, and after the disintegration of the Soviet Union was effectively reversed. The Yeniseysk Daryal radar was lost in 1989 owing to the US protests based on the terms of the ABM treaty. With most of the radar sites outside its own territory, Russia experienced major difficulties in trying to at least preserve the existing network and the coverage it was providing. In a spectacular move, the independent Latvia insisted in 1995 on blowing up the unfinished Daryal radar in Skrunda and declined to extend the agreement allowing Russia's temporary use of the two older radars at a cost of $5 million a year. When the agreement expired on 31 August 1998, SPRN sustained a gap in controlling a critical northwestern sector of its borders.[15]

With the CIS radar-hosting countries the terms of the arrangements were less dramatic but still tense, with the sole exception of Belarus. Ukraine appropriated the radars on its territory for its own military use but shares the data and maintains interoperability with the Russian system. Russia pays rent to Kazakhstan and Azerbaijan for the use of Balkhash and Gabala radar stations. After a decade of disputing with the Azerbaijani government the terms and the very possibility of the use of the Gabala installation, the two countries in January 2002 signed an agreement on the ten-year lease of the station. Pressed by the importance of the Gabala radar for maintaining the 'uninterrupted radiolocation space' and covering the sensitive southern direction, Russia agreed to pay back a controversial debt claimed by Azerbaijan and to pay a $7 million annual rent over the next ten years.[16] The least conflict-prone arrangement exists between Russia and Belarus, according to the 1995 agreement on the 25-year lease of the Baranovichi station land and facilities.[17] The Volga radar near Baranovichi finally became operational in 2002 and restored coverage of the northwestern sector previously only partly provided by the Pechora and Don-2N radars.[18]

The SPRN concept also included the over-the-horizon (OTH) tier capable of detecting US missiles immediately upon their launch before they were visible above the horizon of the Soviet early warning radars. However, this part of the programme was much less successful. Two OTH Duga radars were deployed in the early 1980s at Chernigov, Ukraine ('Duga-1') and Komsomolsk-na-Amure ('Duga-2'). They suffered considerable operational complications and were eventually closed down: Duga-1 in 1986 after the Chernobyl catastrophe and Duga-2 following a fire in 1990.[19]

The early warning network: space-based

The space-based component was meant to be an integral part of the SPRN. In the light of the underperformance of the OTH component, satellites are the only Russian means of detecting a missile attack immediately on launch and thus extending the precious reaction time. The first generation of the early warning satellite system 'Oko' was deployed on highly elliptical orbits (HEO) and was capable of detecting a burning missile motor against the background of space. The scope of the system was deliberately limited to detection of ICBM launches from US territory but not the launches of sea-based missiles. The orbital characteristics allowed a single Oko satellite an observation period of about 6 hours. A continuous 24-hour observation of US territory thus required at least four satellites. The satellite constellation was planned to keep as many as nine HEO satellites, to provide for accuracy and reliability. The system was operational since 1978 and was placed on combat duty in 1982. From 1984, the HEO constellation was supported by satellites in geosynchronous orbit that allow a 24-hour view of any chosen area. Since that time, at least one satellite was deployed at the geostationary point providing the optimal coverage of US territory (known as Prognoz-1 location) almost all the time.[20]

From 1988, the second-generation 'Prognoz' satellites, whose distinguishing characteristic was a 'look-down' capability for spotting missile launches against the Earth's background, started to get placed into geostationary orbit. The Prognoz programme, which had seven dedicated positions reserved with the ITU in 1981, continued through the 1990s but was plagued with increasing failure rates. In 2000 there was a period when the constellation of Prognoz satellites was essentially closed down with no satellites operating.[21] Since then, no more than two second-generation satellites have operated at any one time, the preferred position being the Prognoz-1 geostationary location.[22]

The arrival of the geosynchronous satellites provided a qualitative difference in the reliability of the EW space system, for it ensures its operational ability even in the minimal configuration. Well after it was put on combat duty, the system continued to suffer from complications.

Still, under technical challenges and dwindling financing, the military managed to keep the constellation close to the maximum strength, with eight or nine satellites in HEO orbits and one geostationary satellite, for a remarkable ten years, from mid-1987 through late 1996.[23] Since that year and up to 2002, the space-based early warning system was progressively degrading, with Russia unable to replace the satellites lost to malfunctioning or ageing. This culminated in a fire on 10 May 2001 that destroyed the Serpukhov-15 ground control station and resulted in loss of communication with all four HEO satellites in orbit at that time.[24]

Space tracking

The Soviet Union was developing an elaborate system of space tracking and control, which eventually included networks of stations for tracking of launched missiles and satellites; data-processing, test and command centres;[25] as well as a space surveillance network (*Sistema Kontrolya Kosmicheskogo Prostranstva*, SKKP). Russia continued to develop the inherited assets, with a qualitative change coming around the turn of the century. The latest addition to the space surveillance system was the optical space tracking facility 'Okno' near Nurek in Tajikistan, which in July 2002 was put on test duty and entered combat duty two years later. In 2004 Russia and Tajikistan negotiated an agreement on the terms of the use of the station.[26] The military hailed this development as a sign of the revival of the Russian global strategic capacity.[27]

Back to duty: Russia rediscovers its missile defence potential

Developments in several parallel fields manifest a general trend, slowly materializing at the turn of the century, of a revived state interest in air and missile defence. Facts suggest that both the ABM system and the early warning component of the Russian nuclear forces suffered a loss of the state's attention most badly in the second half of the nineties, with a turning point around the year 1999. One of the first signs suggesting reconsideration of the role of strategic defence was the testing of a Moscow ABM interceptor missile conducted at the Sary Shagan range in the autumn of 1999 after an almost seven-year interval. In spring 2002 a 'planned upgrade' of the Moscow ABM was announced that involved enhancing the early warning back-up by improving the links between the system and the Volga radar at Baranovichi in Belarus made operational around that time.[28] In late 2002 the military conducted another interceptor test and decided it was possible to prolong the ABM service time 'for another several years'.[29] In the same month the newspaper *Izvestia* placed an interview telling about a presidential meeting where participants 'theoretically' discussed the problem of national MD for Russia. With the launch of 'large-scale work' on NMD out of the question, it was decided to resurrect a 'minimal' level of research and development (R&D) on missile defence in order not to lose the scientific and technical potential in the field of MD accumulated over the last four decades. A possibility of modernization of the Moscow ABM 'along the lines of non-nuclear intercept' was floated in experts' comments in the media.[30]

The same trend may be followed in the logic of administrative reorganizations. In this field, the inertia of neglect took a little longer. The lowest point was passed after a four-year period from October 1997 through June 2001 when the Space-Missile Defence Troops (*Voyska PRO i RKO*), previously existing as part of the Air Defence Forces, were merged with the Strategic Rocket Forces (SRF), with a negative effect on their infrastructure and operability. In 2001 the Space-Missile Defence Troops and the Military Space Forces were reorganized into a separate arm of the forces – Space Forces (*Kosmicheskie Voiska*) – to combine the ballistic missile, early warning, space surveillance and military space launch capabilities and ABM and anti-satellite (ASAT) systems.[31] This development reflected recognition of the specific mission and role of these troops. The first Space Forces Commander, Colonel General Anatoliy Perminov, named the task of modernization of the missile attack early warning network as one of the most urgent issues facing his command.[32] According to reports, the Space Forces (and specifically the satellite programmes) enjoyed dramatic increases in funding of 350 per cent between 2001 and 2003, and 210 per cent in the 2004 budget (that may also indicate critically low funding levels prior to 2001).[33] At the beginning of 2003 Commander Perminov told a press conference that he was 'satisfied' with the amount of funding planned for that year. The new Commander, Vladimir Popovkin, appointed in March 2004, represents a dynamic approach to optimization and upgrading of the force, aiming for 'the full reanimation' of the military space in three or four years.[34]

In 2003 and 2004, work was under way to capture the newly discernible military and technological realities into a concept of an integrated all-encompassing system of air and missile defence. In 2003 the military started a controversial reorganization of the air defence, aiming to combine the previously autonomous national and deployed troops air defence systems (PVO) under a joint command.[35] The reorganization is to be based on a unified all-Russia control and command system the extent of whose readiness is as yet unclear. In 2004 Defence Minister Ivanov, without being specific, declared the creation of a new national system of the Air and Space Defence (*Vozdushno-Kosmicheskaya Oborona*). In the press, some speculation followed concerning how the new VKO is going to differ from the already existing systems of PVO and PRO (Missile Defence).[36] In this instance, the terminological overlapping appears to indicate a need for a new conceptual framework and language for the emerging military and technological reality that cuts across the former distinctions, fusing the air and missile defence systems, merging components of strategic and tactical ranges, combining the 'theatre' segments into nation-wide geometries, and even uniting military and civil applications. Reports tell of the government deliberating the concept of the Air and Space Defence (VKO),[37] with no result produced as of early 2005. However, the principles of the future transformation appear to have been generally defined. 'Integration' is going to be the key word for future development of the early warning, tracking and fire capabilities of different ranges into a kind of seamless and multilayered network providing for the optimal coverage. At the present moment, the technical, organizational and financial challenges of such an enterprise appear to be enormous.

Building the air and missile defence: possibilities for international alignment

Strengthening of the air defence tier within the CIS

As defence systems are built on assumptions about probable enemies and allies, the momentous developments in the Russian air and missile defence will both influence and be influenced by patterns of international alignment. At this moment, Russia's alignment preferences display a clear enough but not entirely fixed bent. The post-Soviet space still remains the focus of the security outlook of the Russian military. In the field of missile defence, the geographical spread of the Soviet-built infrastructure puts additional emphasis on relations with the Commonwealth of Independent States (CIS) countries. The main activity, however, is centred on air defence.

As early as 1992 almost all CIS members had signed the Collective Security Treaty, and in 1995 ten CIS nations (Armenia, Belarus, Georgia, Kazakhstan, Kyrgyzstan, Russia, Tajikistan, Turkmenistan, Uzbekistan and Ukraine) created a joint air defence system. This system appears to be one of the few successful CIS institutions, exhibiting high exercise activity and gradually moving to mend the gaps in the air defence over CIS territory. The state of the infrastructure and equipment across the member countries has been gradually made more uniform: in 2000, only the central command posts in Russia, Belarus, Kazakhstan and Ukraine had an automatic mutual information exchange system in operation. Contact with the other central command posts took place via telegraph.[38] In 2001 the Defence Minister, Sergei Ivanov, noted that the CIS members were moving from a unified to an integrated air defence system.[39] In 2003 Armenia, Belarus, Kazakhstan, Kyrgyzstan, Tajikistan and Russia created the Collective Security Treaty Organization (CSTO). Its members can take advantage (starting 1 January 2005) of internal Russian prices and tariffs in transfers of arms and equipment to their countries. In 2004 the air defence forces of Russia, Armenia, Belarus, Kazakhstan, Kyrgyzstan and Uzbekistan carried out joint combat duty, considered the highest form of military cooperation.[40] At the CIS Defence Ministers' Council in December 2003, President Putin called the unified air defence the most effective of the CIS cooperation projects and placed its further development first in the range of cooperation priorities.[41] The Russian defence industry, for its part, has a significant stake in providing the material base for CIS defence integration, with a large market for upgrading the older Soviet-made weapons and equipment and securing new procurement contracts.

Armenia and Belarus traditionally have been closer partners. In 1999 Armenia and Russia formally completed integration of their air defence systems: a joint command centre near Yerevan went on duty and the Russian army unit based in the country was announced to deploy the S-300 anti-aircraft systems.[42] In 2003 Armenia was reported as carrying out its first 'successful' testing of the S-300 during the joint Armenian and Russian exercises held 20 km off the Turkish border.[43]

In September 2003 Defence Minister Ivanov announced an impending lease of the S-300 (*Favorit*) systems to Belarus for defence 'of our common sky' and stated that 'at the moment our military relations with the Belarus brothers are deeper than with any other CIS members'. The press reported a pre-existing work of several years on setting up a 'unified regional air defence system' on the CIS western border. The ERS PVO (*Yedinaya Regional'naya Sistema Protivovozhdushnoi Oborony*) is to combine the air and air defence forces of Belarus, the Russian Baltic Fleet and some units of the similar forces of Russia to provide complete protection of the sky in the western direction over Belarus and Russia capable of resisting 'a most powerful air and missile attack'.

At the same time, closer CIS/CSTO integration in this critically important sphere of defence poses its own problems. An example of possible complications was a June 2004 fall-out between Russia and Kazakhstan, following reports of the Astana contract with the British BAE Systems on modernization of the national air defence. Moscow immediately pointed out that orientation on Western military technologies violates Kazakhstan's obligations under the 1995 agreement on joint air defence.[44] In the end, the reports were declared to be unsubstantiated, their release appearing as an attempt to pressure the Russian authorities for more favourable terms for a similar modernization contracted by the Russian companies. This incident, however, emphasizes two points bearing on the prospect of the CIS-oriented development of the prospective air and missile defence system.

First, the political commitment of the CIS governments warranting Russian investment into security-sensitive infrastructures in the CIS area remains uncertain and dependent on a combination of internal (for each CIS country) and international variables. With the ambitious Russian plan of an all-encompassing sophisticated system embracing functions of air and missile defence, cooperation with the CIS is still concentrated on the air defence end of it. Upgrading the ATBM flank of the CSTO air defence has progressed slowly, with only 'two or three' unspecified modifications of S-300 deployed in Kazakhstan around 2000,[45] a three-year interval before the announced and actual deployment of these systems in Armenia, and several years of finalizing appropriate agreements with the closest ally, Belarus. While the military and defence-industrial sectors of all the countries involved seem to be supporting the prospect of further integration, the political arrangements are not following as smoothly. On the strategic missile defence end, Russia is favouring the principle of autonomy and independence. In 2004 the Russian Space Troops commander reiterated an intention, in the future, to carry out the space launches exclusively from the territory of Russia.[46] Space troops also place hopes, in the coming years, on strengthening the early warning system with a new generation of missile-detecting radars made of 'highly finished' modules and therefore taking minimal time to deploy, which presumably would give more flexibility against a 'stationary' radar network.[47]

Second, not only the CIS governments, but first of all the Russian leadership, still has to define its position vis-à-vis NATO and the US. The military activity of Russia in the CIS, and specifically the development of air defence, has a distinct NATO-centred frame of reference. The Russian policy of strengthening the CSTO as a counterweight to NATO, and boosting the Western direction of the joint air defence system, is inconsistent with declaration of the Russian–NATO partnership at the political level and advancing specific defence and armaments cooperation projects addressed to NATO by the military. The NATO and US military presence in the CIS has to be factored in the Russian strategic outlook in a way consistent with political objectives in other directions. While the 2004 exercises of the joint air defence system were based on a politically neutral scenario of a terrorist threat, the BAE Systems démarche of Astana in the same year immediately evoked the spectre of 'NATO getting access to the CIS joint PVO system', unquestionably implying the adversary nature of the alliance and the centrality of this image to the strategic assumptions of the CIS air defence cooperation. Without managing this inconsistency, long-term defence planning and development in the region will be lacking a firm foundation.

Missile defence: pros and cons of the Western orientation

A European option: theatre missile defence

Around the turn of the century, in parallel with building the CIS air defence system, Russia was probing a Western option in the development of missile defence. In 2001–2 Moscow, in a determined diplomatic offensive, tried to woo Europe into a partnership on non-strategic missile defence. In summer 2000, President Putin during his visit to Italy advanced a surprise offer to build a pan-European missile defence with Russian participation. The proposal was repeated in a more substantive form in February 2001 in a document entitled 'Phases of European Missile Defence',[48] presented by the Russian military to Lord Robertson of NATO, with copies sent to the defence ministries of the UK, Germany, France and Greece. Conceived and articulated at the height of the diplomatic struggle to prevent the US withdrawal from the ABM treaty, the Russian proposal stressed several principal points. First, it emphasized the non-strategic nature of the hypothetical Russian–European system, talking of mobile defence formations to protect deployed forces and, possibly, populated areas, i.e. 'theatre missile defence' (TMD). Next, it put an emphasis on preliminary political consultations on identifying possible threats and the conceptual foundation of the future system rather than on rushing straight into determination of military and technical details, in an implicit allusion to the pushiness of the US approach. Finally, it contained a (hardly practical) stipulation that any regional theatre MD system should not be a 'members only' arrangement but open to all willing participants.

The Russian proposal had originally been addressed to the 'Europeans' at large, and could be connected to the interest Russia displayed towards the emerging European Security and Defence Policy (ESDP) that positioned the European Union as an independent actor in the military sphere. It was owing to the ESDP's low institutionalization, and, inversely, the established ties with NATO, as well as the favourable moment for 'upgrading' the Russian–NATO relationship following 11 September, that the Russian proposal landed in NATO and got a special place among the areas of cooperation chosen for the activities within the Russia–NATO Council established in May 2002. The ad hoc working group devoted to this issue is chaired by Robert Bell who, within NATO, specializes, among other things, in defence investment, armaments cooperation and air defence and has a 25-year background of handling defence policy and arms control in the top echelons of US politics.[49] Russian official statements have been consistently hopeful of the prospects in this area of cooperation.

However, against the background of the military realities, the 2001–2 proposals appear to be no more than political manoeuvring. In its vision of NATO, Moscow perceives a number of highly alarming signs, starting from the first round of the NATO expansion in 1998 and the subsequent intervention in Yugoslavia. The new NATO Strategic Concept adopted in 1999 contained statements ringing direct alarm bells in the Russian military and security establishment. The Concept proclaimed concern with 'instabilities on NATO's periphery [that] could lead to crises or conflicts requiring an Alliance military response, potentially with short warning times', and spelled out a commitment to conduct NATO crisis response operations 'including beyond the Allies' territory'.[50] Analysis of the Western military campaigns of the 1990s and early 2000s has prioritized the danger of the air threat. As stated by the Defence Minister in October 2003, 'the enemy will not arrive in a tank.... It will fly to us in a plane or will deliver the weapons by air.'[51] With the Russian air defence system geared for protection against a hypothetical NATO attack, and the Cold War perceptions still deeply embedded in mental and material structures, the prospect of work on common theatre missile defence, which both in NATO and in Russia follows the concept of extended air defence, cannot appear immediately realizable.

The United States as partner: change of language

The US, with its sweeping BMD programme, stands as the single most important referent for Russian policies in this field. Much of the current Russian reaction to the US programme is predicated on the Soviet superpower heritage of the strategic relationship with the US, including the role accorded in this relationship to the ABM treaty. It was the perceived need to preserve the treaty that initially predetermined the cold shoulder to the US BMD plans. During the several concluding years of the ABM treaty debate, the Russian official line, normally presented by the Ministry of Foreign Affairs (MFA), contrasted the strategic and non-strategic kinds of MD: the strategic MD undermined the strategic stability, and therefore was unacceptable; the non-strategic, theatre MD, both on a bilateral and multilateral basis, on the contrary, was 'stabilizing' and 'genuinely necessary and practicable'.[52] Within this frame of reference, not only cooperation with the US, but development of Russia's own extended missile defence system was out of the question.

However, after the treaty was gone, unilaterally abrogated by the US, the Russian position underwent a change. In December 2002, *Izvestia* carried an almost casual remark by the then Chief of the Russian General Staff Kvashnin on the results of the talks with his US counterpart General Myers, to the effect that Moscow is not against cooperation in building of a 'joint NMD system', with the only provision that each of the sides controls its own 'key'; the information on monitoring the missile launches will be shared, but each side will have its own autonomous fire control centre. The article mentioned the negative attitude of the Defence Ministry, but the bottom line of the piece ran clearly in favour of 'catching up with the leaving train'.[53] One month later, in January 2003, Defence Minister Sergei Ivanov, speaking on the issue of MD cooperation following talks with Japan, started out with the habitual line of the 'two types of MD' but ended up in an unexpected admission that 'theoretically, we do not rule out the possibility of Russia's participation in certain elements of strategic missile defence'.[54] A week later, President Putin also spoke of the possibility of Russia and the United States cooperating in building an anti-missile defence system.[55] Finally, the MFA had moved to modify the earlier position: in a September 2003 interview a highly placed official conceded that 'interaction [with the US] in certain technological directions' was the order of the day.[56]

Thus, at the official level, Moscow signals readiness to consider options of cooperation with the US in missile defence. Yet, for Moscow, such cooperation can only be possible by reaching appropriate political agreements necessary to neutralize the negative effects of the ABM treaty's abrogation and preserve 'predictability, trust and transparency in the interconnected fields of strategic offensive and defensive weapons'.[57] The principle of strategic stability, long invoked in the ABM treaty context, had never been exclusively confined to the sphere of bilateral Russian–US relations in the nuclear strategic field. The concept of strategic stability, in a broad sense, refers to multilateral cooperative security arrangements on the global scale, including international accords and regimes on arms control and disarmament, and nuclear and WMD non-proliferation. Concerned that the individualistic US approach to MD can disrupt mechanisms of regional stability and undermine international multilateral institutions, Moscow has repeatedly stated that it will make any form of Russian substantive participation in the US BMD programmes conditional on reaching a 'new political agreement in the sphere of MD'.[58]

The existing experience of the Russian–US technological cooperation in MD has been far from promising. The only major collaborative project, the Russian–American Observation Satellite (RAMOS), lingered from 1992 through 2004. Designed for testing new ways of detecting missile launches from space, it involved bringing into orbit two satellites equipped with infrared sensors and subsequent joint processing of their data, which could become a part of a larger early warning cooperation. Between 1995 and 1999, Russia and the US performed a series of experiments in the RAMOS framework. The project went through ups and downs in US congressional support, had its timeline extended several times, and was finally terminated in 2004 to be replaced by unspecified 'other' cooperative projects in missile defence.[59]

A parallel project of establishing a Joint Data Exchange Centre located in Moscow to exchange strategic early warning data and advance warning information seems to be faring no better. The two sides agreed on the principle and technicalities of operation and even on minor practical details relating to setting up the facility, but the opening date had been postponed, first in June 2001 and then in May 2002, with no specific date set after this. The implementation has come to a procedural impasse over technical matters, which, however, clearly indicates a shortage of political will on both sides.[60]

Domestic determinants of Russian missile defence policy

Three main contexts of the issue

Decisions on the MD policy involve arguments of different order. The several main contexts in which options are weighed each provide a set of rationales focused on a particular value, with certain established but still not firmly fixed hierarchies.

Hard security considerations come first and, under the present political leadership, are gaining more weight. In the budget for 2004, the respective lines for defence and security are absolute leaders over all other expenditures, together taking a share of about 28.7 per cent against the 13.3 per cent for the combined social programmes. During the three-year period from 2003 to 2005, the trend has been a slow growth in defence and security and a decrease, in relative terms, in the social expenditure part (in 2003, 26.3 per cent against 11.9 per cent; in the proposals for the 2005 budget, 31.3 per cent against 11.5 per cent).[61] Air and missile defence relates directly to the components of military strength considered vital for protection of Russian sovereignty in any international contingency. The latest edition of the Russian military doctrine aired by the Defence Minister in October 2003 emphasizes the need to defend against 'the air and space strikes'. The US and NATO are the principal international actors that not only possess but have massively used these capabilities in the recent period. In relation to the US, the developments in missile defence are linked, if not in the short-term, then in the long-term prospect, with the Russian nuclear deterrent potential, the Russian military in recent years synchronizing their strategic missile tests with the US BMD moves.[62] A large-scale strategic forces exercise was held in February 2004 after more than a 20-year break from similar activities in the Soviet times. In the hard security context, the US and NATO are regarded as opponents much more naturally than 'strategic partners', with only a short break after 9/11.

Considerations of *technological strength*, as a condition of guaranteed defence capability, create the next directly relevant set of rationales. The technological edge of the US, growing with the lavishly funded BMD research, presents as much concern as challenges of a political or military nature. Moscow is aware of its economic vulnerability: the 2000 Concept of National Security notes that the 'state of the national economy' is the primary source of security threats and lists the 'weakening of the national scientific and technological potential' among the most immediate specific concerns.[63] The government is striving to revive the economy with the technological potential of the defence industry, the most favoured economic sector in the Soviet times. In the highly centralized and militarized Soviet economy, military research and development (R&D) concentrated the best of the national material and human scientific resources. Even after a decade of post-Soviet severe underfunding and disintegration, the Russian military

R&D is still admitted to maintain a wide range of leading technologies, in some areas taking positions in the 'absolute forefront'.[64]

Since 2001, budget allocations for technological development have been steadily growing. In 2001 the R&D share in the procurement expenditures (*Gosoboronzakaz*, state defence order) was increased by 43 per cent. For 2003 the procurement funding was reported to be around 110 billion roubles (a 33 per cent increase from the previous year but the same 2.6 per cent of GDP), with a new increase in the R&D share. However, even the increased state funding accounts for no more than 20 per cent of the defence industry R&D expenditure. The remaining 80 per cent comes from the self-financing of the enterprises,[65] i.e. from the sales revenues predominantly coming from exports. The desired technological and industrial revival will not be possible without private investment, including international. Rich in technologies but short of money, Russia regards MD as a possible area for defence-industrial cooperation that could incorporate Russian defence companies into transnational networks of R&D and manufacturing, keep up to the higher standards and thus help escape the technological backwater.

Motives of *prestige* and *status* will also guide decision-making in international matters. Russian doggedness about the preservation of the ABM treaty can in some part be explained by the unwillingness to part with one of the last vestiges of its great power status. In the annual address both in 2003 and 2004, the President stressed that Russia ought to take 'an equal position among the most advanced states'.[66] The theme of the excellence of the Russian military technologies has been a constant currency in Russian media, military and defence-industrial statements. At the top official level, both the President and the Defence Minister have made unequivocal references to the supremacy of the Russian missile and space technologies. Cooperation with the Europeans or the Americans on missile defence may convey a positive value of working with 'the most advanced states' in an important area. Politically, such cooperation may be both a sign of and an instrument for developing a security partnership between Russia and the Western countries.

Actors in the political process

The military

Because of the nature of the issue, the military is in a position to exert direct and indirect influence on the policy. In general, deference to the position of the 'forces' (*silovye struktury*) is a characteristic of the current Russian political system. Sergei Ivanov, the current Defence Minister, is considered among possible 'successors' to the current President.

The military has been traditionally influential in policy-making towards NATO and the US.

On the whole, the military tends to place reliance on the independent Russian air and missile defence capability and distrust collaborative interdependence schemes. At the same time, the military statements on issues of Russian–NATO and Russian–US relations are becoming more guarded and balanced. General Colonel Yuri Baluevsky, appointed in 2004 the Head of the General Staff, was a supporter of the European joint missile defence initiative and reiterates the possibility of Russian–NATO missile defence against 'strategic missiles of terrorist organizations'.[67] Defence Minister Ivanov in 2003–4 repeatedly spoke in support of MD cooperation with the US, with the provision of a 'full state control' over it.[68] The Russian military has participated in a series of staff exercises with the US (and, more recently, with NATO) on theatre missile defence. From another side, the military is usually in favour of selling Russian armaments and equipment abroad, including to non-CIS countries, which presumably would entail support of cooperative co-development or manufacturing projects. On balance, the Defence Ministry and the General Staff are displaying a rhetorical readiness to go ahead with joint missile defence schemes provided the latter are based on a solid political foundation. It is important to remember that the position of the Russian military is strongly conditioned by the policies of their counterparts: extending NATO air defence infrastructure into the territory of the Baltic states or the US upgrade of the Thule and Fylingdales radars (in all cases areas of low priority in terms of terrorist or 'rogue state' threats) understandably evoke Russian suspicions about the role accorded to Russia in NATO or US military planning. The military is more amenable to the partnership discourse on 'neutral' grounds of combating terrorism or rescue operations.

The political scene: a liberal course under a nationalist majority?

On the domestic political arena, an option of security cooperation with the West and, specifically, participation with the US and Europe in the BMD project may still be viable. The Russian political space does contain a tradition of cooperative Western orientation in the security area. Its political representation at present may seem at the lowest possible ebb, but it is still there. The presumption that the achievements of the Russian defence industry merit and in fact require Western investment, and thus encourage Russian–Western military and political cooperation, has long existed on the 'liberal', 'pro-Western' end of the Russian political spectrum. The first Russian pronouncement on record bringing up the prospect of the internationalization of the Russian scientific potential precisely in the context of MD cooperation belongs to the first Russian President, Yeltsin. In July 1992 in a speech at the United Nations General Assembly he proposed 'to deploy and operate jointly a Global System to Protect Against the Threat of Ballistic Missile Attack [GPS], based on a revised American SDI [Strategic Defence Initiative] and advanced technologies developed by the Russian military-industrial complex.'[69] During the final stage of the Bush (senior) administration, the so-called Ross–Mamedov talks were briefly held to discuss some technological and legal issues of a possible joint missile defence.[70] This orientation can be traced into the early 2000s, when it influenced the Russian pan-European TMD initiative, initially put forward by the President but later monopolized by the military.[71]

Since the parliamentary election of December 2003, the Russian political spectrum has grown considerably more monochrome, with the liberal and the left-wing opposition disoriented and seriously weakened. The nationalist discourse perpetuated by the leaders of the basically pro-government Liberal-Democratic Party (LDPR) and the custom-made (for distraction of the traditional communist electorate) block *Rodina*, does not strongly diverge from the 'strong state' ideology of the predominant centrist *Yedinstvo*. Nevertheless, even in the present political situation, accommodation with the US or NATO on MD could be domestically feasible, if two conditions are met. First, if such a move is credibly represented to be in Russian interests and upholds the 'equal' position of Russia vis-à-vis the Western partners. In this case it plays to the vision of a powerful and influential Russia. Second, and more fundamentally, if this course is promoted by the President. As early as May 2000, Putin, barely two months into his presidency, successfully steered the Duma to ratify START-2, the treaty whose ratification had been several times refused by the Russian parliament in the Yeltsin era. Since then, the Russian political system has gone a long way towards a legislature fully responsive to the presidential line. Regarding the President's position on the issue of MD, it is safe to state that he appreciates its significance and is at least aware of the options for and benefits of international cooperation. It was Putin who advanced the initiative of a joint European theatre missile defence. He was also on record speaking of the possibility of cooperation with the US in strategic MD.

Finally, public opinion, notwithstanding the general upsurge of anti-American sentiment in 2002–3, displays a degree of flexibility. Public opinion polls in 2001–3 show that while people tend to hold predominantly negative evaluations of the US role in the world and its attitude to Russia, at the same time they admit and accept the necessity of cooperation.[72]

Important but not enfranchised: the defence industry in the political process

With the Russian defence-industrial interests often invoked as the central rationale for participation in the BMD programmes, the position of the industry itself is not easy to discern clearly. There are a number of factors bearing on the role of the defence industry in Russian policy-making.

First, what is known under the name of OPK (the defence-industrial complex) is a heterogeneous entity composed of bureaucracies, a few large vertically integrated companies combining the entire chain of production, some dozens of economically viable individual research and/or manufacturing companies, and hundreds of enterprises that largely rely on budget financing, a mixture of state and private forms of ownership. Such a conglomerate cannot be expected to form aggregated interests.

Next, even in the segment of the industry specializing in missile defence and composed of comparable units, competition for state funding transcends other interests and precludes possibilities for defining a common position. A third factor that aggravates the situation further is the nature of the Russian political process, which does not provide the defence-industrial entities with mechanisms to openly participate and champion particular projects or strategies. There are few institutionalized channels for representation of the defence-industrial interests in the open political debate and of the public awareness and control over this sector with a rooted culture of secrecy and insulation. This leaves space open for intrigues and backstage deals serving private interests at the expense of corporate and national ones.[73]

Lately, the industry is becoming more proactive in making its interests presented and lobbied in an institutionalized manner. NPO Almaz, the developer of the S-300 and S-400 systems, in April 2003 initiated an 'independent (*vnevedomstvennyi*) expert council on the issues of air and space defence' comprising representatives of the manufacturers, engineers, the military and the government to propose ideas to the national leadership.[74] Since then, the council has advanced a number of proposals, including the concept for development of the future air and non-strategic missile defence system. It should come as no surprise that the prospective system will adopt the Almaz-developed S-400 Triumph as a baseline component,[75] but at least the existence of a body like the council makes the lobbying process more visible and participatory.

In the Russian OPK, there may exist a category of Russian defence companies that would welcome a political accommodation leading to collaborative projects on MD.[76] On the other hand, the attitudes in the industry are in general distrustful concerning the practical matters of scientific and technological cooperation with the US. At the subjective level, there exists 'firm, even subconscious, mistrust of many Russian specialists of the possibility of mutually beneficial and equitable cooperation',[77] confirmed by almost folkloric stories about past episodes of abuse of trust and skimming of the hard-won results of Russian research on the cheap. Officials and industrialists point at instances of the US trying to get samples of the newest Russian weapons for glimpses into the technological inputs, as was the case with the S-300 models.[78]

Conclusion

Summarizing the analysis presented in the chapter, several points should be stressed regarding the contours of the post-ABM transition of Russia's MD philosophy and strategy.

Nuclear (or power) deterrence is going to stay at the heart of Russian defence and security policy, as a guarantee that Russia sustains space and time to conduct internal economic transformation and thus enhance its resources for international action to create a better environment for domestic development. Policy-makers in Moscow realize that effective military power is not going to lose importance in the twenty-first-century world, and its effectiveness will to a large extent rely on technological excellence. In this context, MD will be important for at least four reasons. First, as a factor to be considered for ensuring a long-term strategic stability vis-à-vis the US. Second, as a practical means of providing for security of territory and deployed forces. Third, as an area that manifests and stimulates the nation's technological power. And fourth, as a field (and instrument) of security cooperation and integration.

In setting up a strategy for developing its MD capability, Russia can lay its course between the poles of self-reliance with the state-led resource mobilization on the one hand, and international integration with a higher degree of market-stimulated incentives on the other. The country's limited resources and dissipating technological capacity speak in favour of a cooperative strategy. This strategy is based on the interests of some political and economic groups inside Russia, but it is proving difficult in the sense that it demands innovation and transformation (not only on the Russian part) in some firmly established mental structures and strategic dispositions. So far the Russian military and the state bureaucracies have preserved the Soviet-style world outlook based on the preference for a strong and independent military capability and military integration within the CIS.

However, the CIS scenario does not present the optimal prospects from the point of view of the Russian motivations. On the traditional hard security plane, the CIS orientation to a large extent is in the interest of Russian security. Yet Russia confines military integration in the CIS to the field of air and, increasingly, tactical missile defence, preferring to strengthen autonomy and self-reliance in the space and strategic missile capability. Technologically, the CIS orientation, though attractive in the short-term perspective, is a dead-end for the prospects of Russian defence-industrial and technological development in the long term. Even with a growing Western competition, the CIS markets cannot provide enough incentives to keep abreast with the latest technologies; nor can the

CIS defence industries alone offer internationally competitive and 'upwardly mobile' partnerships for the Russian defence companies. On the other hand, the CIS military integration is going to remain important in terms of status considerations, playing to the inevitable historical memories of the Russian pre-revolutionary and Soviet integral space.

The Western orientation, whether European/NATO or US, is most problematic in the hard security context. Cooperation in missile defence does present an opportunity for a more substantive and therefore more genuine security dialogue with the West. However, the realization of this opportunity is not yet possible under the continuing uncertainty and conflicting swings in the Russian security outlook both at the elite and mass public levels. If Moscow does decide to proceed in the Western direction, the distinction between cooperation with the US, on the one hand, or Europe, on the other, appears to be not as important as the shift in the security relations as such. In any event, no Russian–European, let alone Russian–NATO, cooperative missile defence project can realistically be pursued without, at minimum, the acquiescence of the US, which is the principal BMD actor in the world. From the technological motivation perspective, the Western option opens to the Russian OPK an opportunity to remain in the mainstream of the world scientific and technological development and world-scale standards. This opportunity in no way guarantees that the Russian defence industry will be up to the challenge, but it gives an indispensable stimulus for progressive development that is not available in the CIS direction. Bringing in the status point of view, this hypothetical cooperation will not be free of status-related tensions, with Russia sensitive, as it had previously been, to issues of equality, and the Western partner unavoidably taking the leading role as the one with the most resources. At the same time, substantive partnership in security matters may be regarded as a valuable asset for the Russian international posture.

At the present time, missile defence, or prospects of international cooperation therein, is not an issue of public concern or an object of political contention. Outside of the strategic community that comprises military, industrial and expert individuals, groups and organizations, missile defence is a technical subject of little interest, subsumed in the general contents of relations between Russia and the West. This being the case, the domestic political set-up allows a certain leeway for the support of a cooperative course, should the leadership decide on this course. While appreciating the possible strategic and technological benefits of joining the US BMD programme, Russia gives a greater priority to preservation and strengthening of the multilateral legal frameworks of international governance. Partnership with the US in missile defence will be politically feasible only if the US signals Moscow and the rest of the world a renewed respect for the principles of multilateralism and international legality. As regards the shift in security perceptions that could provide a foundation for the missile defence cooperation between Russia and the West, the policies of the US and NATO, and their strategic disposition towards Russia, acquire a special role in the process of the Russian political and military transformation at the time when Moscow is facing the challenge of setting a strategic course in the security environment of the new century.

Notes

1 Anatoli Antipov and Alexander Manushkin, 'Pro rozhdenie nashei PRO' ('About the Birth of Our ABM'), *Krasnaya Zvezda*, 3 March 2001. Online. Available HTTP: ⟨http://www.redstar.ru/2001/03/03_03/r_or11.html⟩ (accessed 20 August 2004).

2 Hans M. Kristensen, Matthew G. McKinzie and Robert S. Norris, 'The Protection Paradox', *Bulletin of the Atomic Scientists*, 60(2) March/April 2004, pp. 68–77.

3 Pavel Podvig, 'History and the Current Status of the Russian Early Warning System', *Science and Global Security*, 10(1), 2002, pp. 27, 29.

4 'Command and Computing Center 5K80 of the A-135 System.' Online. Available HTTP: ⟨http://pro-pko.narod.ru/a135/kp-a135.htm⟩ (accessed 20 August 2004).

5 Kristensen *et al.* cite several conflicting assessments of the US and British military. A characteristic Pentagon assessment of A-135 points that 'With only 100 interceptor missiles, the system can be saturated, and . . . is highly vulnerable to suppression.' Yet it concludes that '[i]t does provide a defence against a limited attack or accidental launch'. Kristensen *et al.*, op. cit.

6 'Russia: Moscow Shield Obsolete, Former General Says, April 17 2002.' Online. Available HTTP: ⟨http://www.nti.org/d_newswire/issues/thisweek/2002_4_17_misd.html#4⟩ (accessed 20 August 2004).

7 According to an unattributed Web-based estimate, as a result of the blast, around 10 per cent of the population may die, buildings in the range of 177 km^2 be partially destroyed, territory polluted, and all energy and communication lines, including military ones, put out of action. 'Protivoraketnaya Sistema A-135', ('A-135 ABM System'). Online. Available HTTP:⟨http://www.arms.ru/rko/SystemA135.htm⟩ (accessed 12 July 2004). A similar estimate is made by the well-known Russian military expert Vladimir Dvorkin in a recent interview: Dmitrii Litovkin, 'Vladimir Dvorkin: Sozdat' PRO Rossii segodnya neposil'no' ('Vladimir Dvorkin: Building ABM Today is Beyond Russia's Resources'), *Izvestia*, 23 November 2003. Online. Available HTTP: ⟨http://www.izvestia.ru/army/article41370⟩ (accessed 30 July 2004).

8 A careful selection of available statements is provided by an unofficial website 'Voiska Raketno-Kosmicheskoi Oborony', Sistema PRO A-135. Online. Available HTTP: ⟨http://pro-pko.narod.ru/a135/a135.htm⟩ (accessed 15 August 2004). Discussion of the issue is included in Kristensen *et al.*, op. cit.

9 Russia made S-300 (modifications of which are rarely specified) transfers to Kazakhstan and Armenia, and intends to lease them to Belarus. The system was exported to Cyprus (Greece), China and Vietnam.

10 Vladimir Gavrilov and Sergei Ischenko, 'The Triumph Is Nearly Invisible', *CDI Russia Weekly*, 19 August 2003. Online. Available HTTP:⟨http://www.cdi.org/russia/270-15.cfm⟩ (accessed 30 July 2004).

11 'Zenitnye raketnye sistemy "Triumph" nachnut postupat' na vooruzheniye voisk PVO v 2004 godu' ('Air Defence Systems *Triumph* Will be Supplied to the PVO Troops in 2004'), *Interfax-AVN*, 11 March 2004. Online. Available

HTTP: ⟨http://www.raspletin.ru/press-center/articles/interfax-avn_031104_02.php⟩ (accessed 20 August 2004).

12 Vladimir Morozov, 'Vsevidyaschee oko Rossii' ('The All-seeing Eye of Russia'), *Nezavisimoye voyennoye obozreniye*, 14 April 2000. Online. Available HTTP: ⟨http://nvo.ng.ru/wars/2000-04-14/4_sprn.html⟩ (accessed 4 September 2004).

13 The article of Pavel Podvig (see note 3) has several graphs illustrating historical stages of the extent of Soviet radar coverage.

14 Ibid.; Morozov, op. cit.

15 Felix Gormley, 'Russian Radar Crisis', *Jane's Intelligence Review*, 1 May 1998.

16 V. Mokhov, 'Rossiya arendovala RLS "Daryal" v Azerbaijane' ('Russia Has Leased the *Daryal* Radar Station in Azerbaijan'), *Novosti Kosmonavtiki*, 3, 2002. Online. Available HTTP: ⟨http://www.novosti-kosmonavtiki.ru/cotent/numbers /230/34.shtml⟩ (accessed 2 June 2004).

17 Alexander Ovchinnikov, 'V polozhenii slepogo boksera' ('In a Blind Boxer Position'), *Nezavisimaya Gazeta*, 18 February 2002. Online. Available HTTP: ⟨http://www.ng.ru/sodr/2002-02-18/7_boxer.html⟩ (accessed 4 September 2004).

18 'Radar System to Guard Northwest', *The Russia Journal* / RBC, 2 October 2000. Online. Available HTTP: ⟨http://www.therussiajournal.com⟩ (accessed 20 August 2004).

19 Michael Jasinski, 'Russia: Strategic Early Warning, Command and Control, and Missile Defence Overview', Nuclear Threat Initiative website. Online. Available HTTP: ⟨http://www.nti.org/db/nisprofs/russia/weapons/abmc3/c3abmovr.htm⟩ (accessed 20 August 2004); 'Osnovnye Vekhi Kosmicheskih Voisk' ('Main Stages of the Space Forces'). Online. Available HTTP: ⟨http://www.old.mil.ru/index.php?menu_id=69⟩ (accessed 20 August 2004).

20 Podvig, op. cit.; Phillip S. Clark, 'Russia's Geosynchronous Early Warning Satellite Program', *Jane's Intelligence Review*, 1 February 1994.

21 Phillip S. Clark, 'Decline of the Russian Early Warning Satellite Programme', *Jane's Intelligence Review*, 1 January 2001.

22 Philip S. Clark, 'Russia Begins to Expand Early Warning Satellite Network', *Jane's Defence Weekly*, 17 April 2002.

23 Podvig, op. cit., p. 49.

24 Philip S. Clark, 'Fire Cuts Link with Russian Satellites', *Jane's Defence Weekly*, 16 May 2001.

25 'Soviet Space Tracking Systems', *Encyclopedia Astronautica*. Online. Available HTTP: ⟨http://www.astronautix.com/articles/sovstems.htm⟩ (accessed 2 September 2004); 'Kosmicheskiye Voiska Rossii' ('Russian Space Forces'). Online. Available HTTP: ⟨http://www.novosti-kosmonavtiki.ru/content/numbers/243/30.shtml⟩ (accessed 2 September 2004).

26 Yurii Gavrilov, 'Kosmos tzveta khaki' ('Space in Colour of Khaki'), *Rossiiskaya Gazeta*, 15 July 2004. Online. Available HTTP: ⟨http://www.rg.ru/2004/07/15/kosmos.html⟩ (accessed 15 July 2004).

27 See a selection of materials in 'Okno v kosmos' ('A Window into Space'). Online. Available HTTP: ⟨http.//www.astronomer.ru/news.php?action=1&nid=90> (accessed 2 September 2004).

28 David C. Isby, 'Russia Plans Moscow ABM Upgrade', *Jane's Missiles and Rockets*, 1 April 2002.

29 David C. Isby, 'Russia Tests ABM Interceptor', *Jane's Missiles and Rockets*, 1 December 2002.

30 Dmitri Litovkin, 'Nuzhna li Rossii PRO i kak yeyo razvivat' ('Does Russia Need a NMD and How to Develop It'). Interview with Vladimir Dvorkin.

Online. Available HTTP: ⟨http://www.inauka.ru/technology/article37527.html⟩ (accessed 15 July 2004).

31 Jasinski, op. cit.

32 However, a further reorganization, with a view to bringing together missile and air defence, looms with yet unclear prospects amid unproductive bureaucratic clashes. Alexander Babakin, 'Kosmicheskaya oborona kolebletsya' ('The Space Defence is Undecided'), *Nezavisimaya Gazeta*, 19 December 2003. Online. Available HTTP: ⟨http://nvo.ng.ru/forces/2003-12-19/1_cosmos.html⟩ (accessed 4 September 2004).

33 Henry Ivanov, 'Quality, Not Quantity', *Jane's Defence Weekly*, 17 December 2003; Vladimir Mukhin, 'Gosoboronzakaz budet prinyat do kontza etogo goda' ('State Defence Order Will Be Approved Before the Year's End), *Nezavisimoye Voennoye Obozreniye*, 26 December 2003. Online. Available HTTP: ⟨http://www.ng.ru/economics/2003-12-24/3_goz.html⟩ (accessed 4 September 2004).

34 Perminov, cited in Vyacheslav Danilenko and Alexander Dolinin, 'U etih voisk i zadachi kosmicheskiye' ('These Forces Aspire to High Tasks'), *Krasnaya Zvezda*, 31 January 2003; Popovkin, in Gavrilov, op. cit.

35 *Nezavisimoye Voennoe Obozreniye* gives space to opposing opinions on the air defence reorganization. For an example of a critical assessment, see Ivan Yerokhin, 'Reorganizatzionnyi razval' ('Reorganizational Destruction'), *Nezavisimoye Voennoe Obozreniye*, 28 May 2004. Online. Available HTTP: ⟨http://nvo.ng.ru/concepts/2004-05-28/4_sistem.html⟩ (accessed 15 June 2004).

36 'Kosmicheskaya krysha' ('A Space Roof'), *Vremya Novostei*, 26 March 2004. Online. Available HTTP: ⟨http://www.vremya.ru/2004/50/4/94770.html⟩ (accessed 2 June 2004).

37 Commander of the Space Troops Popovkin mentioned in a July 2004 interview: 'As far as I know, the concept (*kontzeptziya*) of VKO has not yet been approved and is under deliberation.' Gavrilov, op. cit.

38 'CIS Develops Joint Air Defence System', *Nezavisimaya Gazeta*, 11 September 2000.

39 Lyuba Pronina, 'CIS Conducts Air-Defence Tests'. Online. Available HTTP: ⟨http://www.avia.ru/english/articles/doc65.shtml⟩ (accessed 15 June 2004).

40 *Interfax-AVN*, 12 February 2004.

41 V. Putin, 'Voennoye sotrudnichestvo – vazhneishii faktor vzaimodeistviya na postsovetskom prostranstve' ('Military Cooperation – Most Important Factor of Interaction in the Post-Soviet Space'), *Novosti VPK i Voenno-Technicheskogo Sotrudnichestva*, 6–11 December 2003. Online. Available HTTP: ⟨http://www.mfit.ru/defensive/obzor/ob11-12-03-4.html⟩ (accessed 15 July 2004).

42 Emil Danielyan, 'CIS Defence Chiefs Map out Accord Despite Divisions', *Asia Times Online*, 25 May 1999. Online. Available HTTP: ⟨http://www.atimes.com/c-asia/AE25Ag01.html⟩ (accessed 4 September 2004).

43 'Armeniya ispytala kompleksy S-300 na "voobrazhaemom agressore"' ('Armenia Has Tested the S-300 Systems on the "Imaginary Aggressor"'), 12 December 2003. Online. Available HTTP: ⟨http://main.izv.info/world/12-09-03/news59572⟩ (accessed 15 July 2004).

44 Gennadii Sysoev, 'Vozdushnaya Vstrevozhennost' ('Air Alert'), *Kommersant*, 15 June 2004. Online. Available HTTP: ⟨http://www.kub.kz/article.php?sid=6173⟩ (accessed 4 September 2004).

45 Marat Kenzhetaev, 'Kazakhstan's Military-Technical Cooperation with Foreign States: Current Status, Structure and Prospects.' Online. Available HTTP: ⟨http://mdb.cast.ru/mdb/1-2002/at/kmtcfs/⟩ (accessed 2 September 2004).

46 'Russia to Make All Space Launches From Own Territory in Future', 3 April 2004. Online. Available HTTP: ⟨http://www.rednova.com/news/stories/1/2004/04/03/story104.html⟩ (accessed 15 July 2004).

47 'Novoye pokoleniye radarov rezko povysit effektivnost' sistemy rannego preduprezhdeniya' ('New Generation of Radars Will Strongly Increase Performance of the Early Warning System'), *Izvestia*, 28 July 2004. Online. Available HTTP: ⟨http://news.izvestia.ru/community/news87043⟩ (accessed 15 July 2004).

48 'Zamysel i etapy sozdania obsheevropeiskoi sistemy protivoraketnoi oborony' ('Design and Stages of Building the All-European Missile Defence'). Online. Available HTTP: ⟨http://www.armscontrol.ru/start/rus/docs/evropro.htm⟩ (accessed 2 September 2004).

49 Robert Gregory Bell, 'Who Is Who at NATO?' Online. Available HTTP: ⟨http://www.nato.int/cv/is/asg-ds/bell-e.htm⟩ (accessed 10 June 2004).

50 The NATO Strategic Concept of 1999 also contains a direct reference to '[t]he existence of powerful nuclear forces outside the Alliance', a passage sensitive for Russia.

51 Cited from Viktor Litovkin, 'Doktrina navyrost' ('The Doctrine to Grow'), *Problemy Global'noi Bezopasnosti*, N 13, Online. Available HTTP:⟨http://www.cwpj.org/Publications/gsi/n13/2.htm⟩ (accessed 4 June 2004).

52 See statement of the Russian Ministry of Foreign Affairs, 'On Actvization of the US Efforts to Create a Global Missile Defence', 18 December 2002, and the response of the MFA official representative A.V. Yakovenko to a question from the Russian media about involvement of other countries in the US work on building the global MD, 24 December 2002.

53 Dmitri Litovkin, 'Odna na vsekh. No porozn' ('For Common But Separate Use'), *Izvestia*, 18 December 2002. Online. Available HTTP: ⟨http://izvestia.ru/politic/article27920⟩ (accessed 15 July 2004).

54 'Russia Offers Japan Common Missile Defence', 15 January 2003. Online. Available HTTP: ⟨http://www.rusnet.nl/news/2003/01/15/politics03.shtml⟩ (accessed 20 August 2004).

55 'Russia Offers US to Build Common Missile Defence', 24 January 2003. Online. Available HTTP: ⟨http://www.rusnet.nl/litics04.shtml⟩ (accessed 20 August 2004).

56 Interview with the Deputy Foreign Minister Sergei Kislyak by the newspaper *Vremya Novostei*, published 24 September 2003. Online. Available HTTP: ⟨http://www.ln.mid.ru/Bl.nsf/0/459E0C94F8EA48BC43256DAB002E57C1?OpenDocument⟩ (accessed 20 August 2004).

57 The response of the MFA official representative A.V. Yakovenko, 24 December 2002.

58 Ibid.

59 Nikolai Novichkov, 'Tumannye perspektivy sotrudnichestva' ('Murky Prospects for Cooperation'), *Nezavisimoye Voennoye Obozreniye*, 12 February 2004. Online. Available HTTP: ⟨http://www.ng.ru/nvo/2004-02-13/10_perspective.html⟩ (accessed 20 August 2004); Anneli Nerman, 'Weldon Speaks of Boosting Missile Defence Cooperation with Moscow.' Online. Available HTTP: http://www.centredaily.com/mld/centredaily/news/politics/8755770.htm⟩ (accessed 20 August 2004). On the problems of the project, see Pavel Podvig, 'US–Russian Cooperation in Missile Defence: Is It Really Possible?', *PONARS Policy Memo Series*, April 2003. Online. Available HTTP: ⟨http://www.russianforces.org/podvig/eng/publications/misc/20030425ponars.shtm⟩ (accessed 20 August 2004).

60 'Joint Data Exchange Centre Delayed', 11 June 2001. Online. Available HTTP: ⟨http://www.nti.org/db/nisprofs/russia/weapons/abmc3/stratc3.htm⟩ (accessed 20 August 2004).

61 Svetlana Samoilova, 'Populyarnyi budget nepopulyarnyh reform' ('A Popular Budget of Unpopular Reforms'), Politcom.ru, 21 June 2004. Online. Available HTTP: ⟨http://www.politcom.ru/2004/zloba4317.php⟩ (accessed 20 August 2004); Andrei Sedov, 'Budget-2004: Gotovimsya k voine?' ('Budget-2004: Preparing for War?'), *Komsomolskaya Pravda*, 2 December 2003. Online. Available HTTP: ⟨http://www.kp.ru/daily/23169/25088/⟩ (accessed 20 August 2004).

62 Four days before Clinton's Deputy Secretary of State Strobe Talbott arrived in Moscow for talks on amendments to the ABM treaty in September 1999, the Russian military launched the latest Topol-M ICBM, which had hit its target in the Russian Far East 'with a high degree of accuracy'. Jonathan Steele (*The Guardian*), 'US Bid to Alter Arms Treaty Alarms Russia's Military', *Johnson's Russia List* 3493, 11 September 1999. Online. Available HTTP: ⟨http://www.cdi.org/russia/johnson/3493. html ##2⟩ (accessed 20 August 2004).

63 Concept of National Security of the Russian Federation', *Nezavisimoye Voyennoye Obozreniye*, No. 1 (174), 14 January 2000.

64 Jan Leijonhielm, Jenny Clevström, Per-Olov Nilsson and Wilhelm Unge, 'Russian Military-Technological Capacity' (abstract and executive summary), 2003. Online. Available HTTP: ⟨http://www.foi.se/english/index.html⟩ (accessed 20 August 2004).

65 Ibid.

66 See 'Poslaniye Prezidenta Rossii Federal'nomu sobraniyu' ('Address of President of Russia to the Federal Assembly'), 2003 and 2004. Online. Available HTTP: ⟨http://www.vesti.ru/files.html?id=3924&tid=15307⟩ (accessed 20 August 2004).

67 Vladimir Ivanov, 'General-polkovnik Baluyevsky – Post prinyal!' ('General Colonel Baluyevsky –New Appointment'), *Nezavisimoye Voennoye Obozreniye*, 23 April 2004. Online. Available HTTP: ⟨http://nvo.ng.ru/forces/2004-07-23/3_baluevskiy.html⟩ (accessed 20 August 2004).

68 Dmitry Andreev, 'Proverka svyazi' ('Communication Test'), *Vremya Novostei*, 16 August 2004. Online. Available HTTP: ⟨http://www.vremya.ru/2004/145/4/105226.html⟩ (accessed 20 August 2004).

69 Cited from Andrei Shoumikhin *et al.*, 'Evolving Russian Perspectives on Missile Defence: The Emerging Accommodation', Fairfax, Va.: National Institute for Public Policy, 2002, p. 1.

70 At least one official MFA statement at that time was made on the irrelevance of the ABM treaty to this specific partnership: the treaty was said to restrict 'national defences, while a global defence system that is to be developed, created and operated jointly on a multilateral basis, is not viewed by us as a national system'. Foreign Ministry spokesman on 17 July 1992, cited in Shoumikhin, op. cit., p. 2.

71 For the political background of the all-European TMD initiative, see Alla Kassianova, 'Russian–European Cooperation on Theatre Missile Defence: Russian Hopes and the European Transatlantic Experience', *Nonproliferation Review*, 10(3), 2003, pp. 71–83.

72 See the database of the *Fond Obschestvennoe Mnenie* (Public Opinion Foundation) at www.fom.ru. Most of the Russian people in September 2003 believed that relations between Russia and the US are best characterized as those of 'reluctant partners': see the poll of the Russian population on the eve of Putin's visit to the US, 11 September 2003. Online. Available

HTTP:⟨http://bd.fom.ru/report/cat/frontier/rossiya_i_stran_mira/truck_West/Russia_USA/ra030910⟩ (accessed 20 August 2004).

73 As described in the case of the struggle between two well-established contractors in the area of MD, the Interstate Stock Corporation Vympel (formerly Central Production and Scientific Association 'Vympel') and the joint stock company RTI Systems to obtain contracts for maintenance and upgrading of the Moscow ABM. See Pyotr Polkovnikov, 'Okno uyazvimosti' ('The Window of Vulnerability'), *Nezavisimoye Voennoye Obozreniye*, 24 February 2003. Online. Available HTTP: ⟨http://nvo.ng.ru/armament/2003-02-21/1_window.html⟩ (accessed 20 August 2004).

74 'Sozdayetsya vnevedomstvennyi ekspertnyi sovet po kompleksnoi probleme "Vozdushno-kosmicheskaya oborona"' ('An Independent Expert Council on the Complex Issue of "Air and Space Defence" is Being Created'), *Interfax-AVN*, 11 April 2003. Online. Available HTTP: ⟨http://www.raspletin.ru/press-center/articles/interfax-avn_030411_01.php⟩ (accessed 20 August 2004).

75 'V Rossii k 2012 godu mozhet byt' sozdana novaya sistema zenitnoi oborony PVO i nestrategicheskoi PRO' ('By 2012, Russia May Build a New System of Air Defence and Non-Strategic Air Defence'), *Interfax-AVN*, 27 January 2004. Online. Available HTTP: ⟨http://www.raspletin.ru/press-center/articles/interfax_040127_01.php⟩ (accessed 20 August 2004).

76 In 2002–3, following the ABM treaty demise, there were tentative contacts between Lockheed Martin and its long-time Russian partner in space projects, the Khrunichev Space Centre, as well as between Boeing and RTI Systems to explore possibilities of cooperation on MD. *Jane's Defence Industry*, Mergers, Acquisitions and Teamings, The Boeing Company, 1 September 2003.

77 'Russia–US ABM Cooperation Budding', *CDI Russia Weekly*, ⟨http://www.cdi.org/russia/239-5.cfm⟩ (accessed 4 September 2004).

78 A recent instance is a comment by a top official from the State Committee for Foreign Military and Technical Cooperation: 'Before, and as well now, our American partners were trying to get hold of our equipment samples.... When we were proposing a more equitable cooperation, all we used to get was a polite refusal.' Novichkov, op. cit.

5

RELUCTANT ALLIES?

Europe and missile defence

Sten Rynning

Introduction

European reactions to US plans for a missile defence system vacillate along with the plans themselves, although one might tempt the conclusion that European reactions in the past were predominantly negative but are now slowly changing. Deterrence and mutual vulnerability have made up the foundation of Europe's long peace since 1945 and it is perhaps not surprising that European governments have been reluctant to abandon a blueprint for security that once shocked because of its tantalizing content – we should expose ourselves to nuclear strikes – but which turned out to do the job, that is, provide security.

The principal reason for abandoning deterrence as an overall strategic framework must be linked to the nature of threats: if they cannot be deterred, then we must deny the enemy opportunities, which is to say defend ourselves and possibly take simultaneous offensive action. The problem for the US and its European allies, and thus for NATO as a collective alliance, is that threats are ambiguous and nourish contradicting claims. Are leaders of rogue nations developing the missiles against which the US wishes to defend itself rational and thus deterable, or are they motivated by ideology or religion to such an extent that they blind themselves to our counter-threats, effectively undermining the rationality of deterrence? Some might argue that states are by definition deterable because they can be located and destroyed – they cannot run and hide – which then leads to the next question of whether new actors that do not possess fixed territories and structures of government such as terrorist networks can be deterred? Maybe not, but is the likelihood of their gaining access not only to weapons of mass destruction but also missile technology large enough to justify missile defence?

European allies have mostly answered in the negative, tending to see more continuity than change in their ability to deter threats. But impetus for political change has made an impact. The stronger the US plans for missile defence, the greater the consideration for it in Europe because of the wish to maintain strong transatlantic links. Moreover, the stronger the presence of new threats – notably in the wake of the 11 September 2001 terrorist attacks – the greater the European willingness to question the past assumption that enemies can be rationally deterred. To the extent that these two political processes reinforce each other, the European allies become more willing to adopt missile defence.

Viewed from a bird's-eye perspective, the European allies have moved from a position of reluctance to one of careful endorsement. But it is not a uniform movement. Continued reluctance is nourished by the fact that the two processes of US policy and new threats do not always reinforce each other: for instance, the George W. Bush administration came into office in early 2001 with a clear focus on rogue nations and missile defence, but the terrorist attacks of 11 September that same year robbed the focus of some of its justification and caused policy attention to shift.

This chapter will first take stock of the European position on the issue of missile defence and then explain how and why it evolved as it did. The first section will examine predominantly NATO but also national policy to pinpoint the extent to which a common European position can be said to exist. The subsequent sections will then explain how it came about, moving through distinct phases and relating the trajectory of the European position to the two political processes of predominant threats and US policy. The last section will outline future trends.

Post-Prague: Europe supports missile defence

European allies in NATO support a collective policy position that closely resembles that outlined in the US national security doctrine, raising the question of whether transatlantic harmony is prevailing after a period of tension. We turn first to the policy position, then to evidence of continuing tension.

The NATO heads of state and government defined NATO's position on new threats when they met in Prague in November 2002. They announced a number of decisions which, they further declared, 'provide for balanced and effective capabilities within the Alliance so that NATO can better carry out the full range of its missions and respond collectively to those challenges, including the threat posed by terrorism and by the proliferation of weapons of mass destruction and their means of delivery'.[1] This broad threat assessment clearly includes the missile threat, although in indirect terms, and among the specific initiatives that followed from the assessment were two of which we should take note.[2]

First, NATO endorsed a number of defence programmes aimed at enabling a defence against weapons of mass destruction – supposedly the type of munitions with which rogue states would equip their missiles. The programmes involve the establishment of a deployable NBC Analytical Laboratory and a NBC Event Response team – both in prototypes – along with a centre for excellence for NBC Weapons Defence, a defence stockpile and surveillance system. The accent is thus placed on the gathering of information, including in the case of an attack that may involve WMD. Other defence programmes that relate to the equipment of forces fall within the regular NATO defence planning process and the goals established for it, which brings us to the second initiative.

NATO in Prague 'initiated a new NATO Missile Defence feasibility study to examine options for protecting Alliance territory, forces and population centres against the full range of missile threats, which we will continue to assess'. Feasibility is a complicated matter, however, involving control and command issues in addition to performance and cost, and NATO's study will be concluded only in 2005. The first phase of the study, the selection of a contractor for carrying out the study on NATO requirements, was completed on schedule in September–October 2003 when NATO chose a transatlantic consortium involving, among others, Boeing and EADS. The consortium has 18 months to complete the study, after which NATO will enter a decision-making phase.[3]

No decision has been made, then, but we know that the issue continues to preoccupy the Alliance. For instance, when NATO defence ministers endorsed the feasibility study at their informal meeting in October 2003, in Colorado, they also engaged in a 'crisis management exercise' that involved an evacuation situation far from NATO territories, which was simple enough, but with the added complication that cruise and ballistic missile threats were introduced (coming from 'a freighter-type ship sitting off the coast').[4] Moreover, the NATO summit in Istanbul in June 2004 reiterated the conclusions from Prague: 'Terrorism and the proliferation of Weapons of Mass Destruction (WMD) and their means of delivery currently pose key threats and challenges to Alliance and international security.'[5]

The decision to award a contract, and the fact that it was awarded on schedule as well as being funded through NATO's common Security Investment Programme, reflect growing support for the idea that if missile defence is feasible it should be acquired. Moreover, the political approval of missile defence at this strategic level complements earlier decisions to initiate NATO studies of so-called theatre missile defence where the main issue of concern is the protection of deployed forces.

Setting up *ad hoc* study groups in 1996 to address theatre missile defence, NATO in 1998 decided to formalize the commitment to study the issue by establishing a TMD Study Group, and then in June 2001 by awarding feasibility study contracts to two consortia.[6] With the combined commitment to investigate the feasibility of tactical as well as strategic missile defence, NATO is preparing to adopt an integrated approach to the issue involving – in theory, at least – a seamless system of defences against all types of missiles.

The focus here is on strategic missile defence even though the tactical and strategic can be difficult to disassociate. The issue is, simply, a global one, as US Secretary of Defence Rumsfeld has remarked on several occasions, providing the illustration that threats that are tactical for the US may be strategic for a US ally like Japan. Still, the tactical and strategic ought to be kept distinct. The distinction used to be important because the US and Russia needed a demarcation line separating TMD, which they both pursued, and MD, which was proscribed by the Anti-Ballistic Missile (ABM) treaty and which Russia opposed.[7] The ABM treaty is now defunct, but the tactical/strategic distinction continues to be important because it reflects a common line of thinking among all countries involved in missile defence, namely a readiness to accept missile defence at lower, tactical levels and a different, often more reluctant attitude to missile defence questions relating to territorial defence. In Europe, the principal players are all involved in theatre missile defence developments (see Figure 5.1) but they have to date been critical of strategic defences.

It is possible that the European allies, in contrast to the recent past, are ready to couple existing TMD to strategic capabilities in a multilayered system capable of protecting deployed troops as well as national homelands. If Poland is a yardstick to judge by, the allies are ready. The Polish government proposed in mid-2002 that the US could construct 'a long-range radar station, aimed at monitoring ballistic missile threats from the southeast, on Polish soil' and also indicated that Polish participation in American plans 'should go far beyond simple ballistic missile monitoring'.[8] In July 2004, Poland along with the Czech Republic indicated that they were willing to consider hosting ground installations not only for radars but also the ground-based interception station that the Bush administration might construct in addition to the two stations already under construction in the US (in Alaska and California).[9]

Poland may not be an appropriate yardstick, however. NATO as a whole is only at the stage of studying feasibility and, as the Istanbul summit of 2004 made clear, NATO is only at the stage where 'missile threats' must be dealt with through 'an appropriate mix of political and defence efforts, along with deterrence'.[10] The decision to invest in strategic defence capabilities and deploy them is likely to be difficult, therefore. The following analysis traces European reactions to American missile plans and regional security questions in several phases, thus preparing the ground for an overall assessment of the depth of support

Great Britain	Participated in the November 2003 decision to procure a range of Aster theatre defence missiles, along with France and Italy. The British plan is to fit the defences on its Daring class of destroyers. This is the first British TMD capability.
France and Italy	Have jointly developed a national capability in the shape of the Aster missile system, the first generation of which consisted of a naval and land component (naval and land SAAM respectively). Italy was financing 17 per cent of the transatlantic MEADS programme, which encountered difficulties, however, and which has thus been merged with the Patriot/PAC-3 programme. Italy is continuing its participation but on terms that remain subject to negotiation.
Germany	Financed 28 per cent of the MEADS programme but is also now involved, along with Italy and the US, in the production of a combined PAC-3/MEADS capability. Germany has committed itself to upgrade from Patriot to PAC-3 although, as in the case of Italy, the terms of participation are subject to negotiation.
Spain	Has bought the American Aegis defence system for its frigates but has not yet decided to acquire the missiles that would upgrade the system to TMD levels.
The Netherlands	Has Patriot capabilities and has requested sales of PAC-3 for an upgrade. The Netherlands has in addition become the lead nation in the European Union's capability group charged with studying TMD options for the European rapid reaction force.

Figure 5.1 Theatre missile defence in Europe.

for NATO's Prague commitments and the likelihood that the support will continue.

The Clinton presidency: European anxiety and resistance

Europeans sceptical of missile defence generally found comfort in the Clinton presidency's reluctant engagement in the issue, but by mid-1999 needed to cope with a new consensus within the US that missile defence had to be developed. The consensus came about as the Clinton administration sponsored a Missile Defense Act that declared 'it to be the policy of the United States to deploy a national missile defense'.[11] The act contained language that satisfied both sceptics and proponents in the US, making deployment contingent on technical feasibility as well as US-Russian arms control, and also calling for merely a limited defence system. Moreover, subsequent to the Congress's adoption of the act in March 1999, President

Clinton outlined four conditions for deployment the following July when he signed the act: necessity by threat, feasibility in terms of both cost and operation, and adherence to a renegotiated ABM treaty. These conditions were comforting to European sceptics, but still, this was just the third occasion on which political opponents in Washington had come together on the missile defence issue.[12]

The European position continued to be non-existent, however, at least in substantial terms. European governments generally shared a reluctance to endorse the plans but they differed in terms of threat perception and strategic conception, some believing that the threat simply was not there, others that threats were indeed present but could be countered through strategies of deterrence. In consequence, the European pillar in the Atlantic Alliance was marked by varying political strategies – some governments choosing to voice reservations, others to hedge their bets – as well as ambivalent support for US leadership in NATO. This state of European affairs frustrated US policy-makers as much as it offered them an opportunity to argue that Alliance affairs could not be properly conducted without firm US leadership.

Instrumental to the crafting of a political consensus in the US was Donald Rumsfeld, who headed a bipartisan commission on missile defence that began its work in January 1998, and which had come about owing to the frustration of Congress Republicans with the gap existing at the time between their own perceived need for missile defence and the assessment of the US intelligence community that a missile threat was at least 15 years away.[13] The Rumsfeld Commission reported its conclusions in July 1998 and managed, to the surprise of many observers, to reach consensus. The Commission found that:

- 'The newer ballistic missile-equipped nations' capabilities will not match those of US systems for accuracy or reliability. However, they would be able to inflict major destruction on the US within about five years of a decision to acquire such a capability.'
- 'The threat to the US posed by these emerging capabilities is broader, more mature and evolving more rapidly than has been reported in estimates and reports by the Intelligence Community.'
- 'The US might well have little or no warning before operational deployment.'[14]

The report's conclusions were pointed, and external developments conspired to thrust the report into the public arena. The first such development occurred in April–May 1998 when Pakistan tested its Ghauri medium-range missile and then went on to conduct underground nuclear tests, in a direct response to similar Indian tests, that proved what the world had strongly suspected, that the two countries were nuclear powers. Proliferation and missile tests were thus on the agenda before July, when the next development took place in Iran where the Shahab 3 missile with a range of 1,300 km was tested just as the Rumsfeld report was published.[15] Finally, in August, North Korea tested a three-stage missile (the Taepo Dong I), which, although the test ultimately failed, surprised observers by revealing North Korea's near-ability to put satellites – however small and simple – into orbit.

While external developments seemed to confirm the heightened awareness of a missile threat, other developments blunted the report's conclusions and made it possible for European allies to sustain their policy of scepticism. First, the Rumsfeld Commission, while bipartisan in composition, was presented by some as a typical product of partisan thinking – essentially, longstanding Republican ideas – and it competed for attention with more sceptical reports from the General Accounting Office and the Central Intelligence Agency.[16] Secondly, the Rumsfeld Commission was criticized for adopting a standard of evaluation involving what 'could' happen rather than what was 'likely' to happen. The CIA thus conceded in the wake of the report, notably in light of the North Korean missile test, that a missile attack 'could' happen earlier than foreseen but continued to insist that such an attack was the least 'likely' faced by the US and that terrorism figured as the top priority in their threat assessment.[17]

A gap was opening in the US position that allowed European governments to continue rather than change their missile defence policies and instead focus resources on the conflicts that erupted or threatened to erupt in their near abroad. The conflict drawing the most attention from European governments at this time, 1998–9, was the Serb–Albanian clash in Kosovo, which ended with NATO's armed intervention in the spring of 1999. Kosovo-type conflicts and crisis management more generally also became the inspiration for the European Union's security and defence policy (ESDP) which was introduced and launched in 1999. The stability of Russia and its continued transformation into a democratic and reliable partner represented another issue of concern, notably on the grounds that Russia needed to be drawn into an extensive web of cooperation that included arms control agreements, such as the ABM treaty, and which moved the focus from military to political modernization. Moving beyond Europe's near abroad, one could also point to China's role in achieving the stable, multilateral world order that European governments generally promote.

Kosovo, Russia and China were the main sources of European reluctance to prioritize missile defence in 1998–9 when President Clinton aligned with Republicans and signed the Missile Defense Act. European governments ended up making concessions to the US within NATO but only after having protested, and in spite of NATO's compromise on strategic policy the protests continued, as we shall see below.

Fragile NATO consensus

European concessions are found in NATO's Strategic Concept which was adopted in April 1999 and which replaced the former concept from 1991. The concept directs the Alliance's political and military planning, and paragraph 56 of the concept directs planning to take into account missile defence:

> The Alliance's defence posture against the risks and potential threats of the proliferation of NBC weapons and their means of delivery must continue to be improved, including through work on missile defences. As NATO forces may be called upon to operate beyond NATO's borders, capabilities for dealing with proliferation risks must be flexible, mobile, rapidly deployable and sustainable.[18]

The door opened for both strategic and tactical missile defence, although the latter received greater attention given the underlying consensus that forces needed missile defence in their area of operation, especially now that all agreed that NATO would go 'out of area' to avoid going 'out of business'. Thus, the aforementioned TMD Study Group from 1998 was boosted by the decision to involve TMD in the Defence Capability Initiative adopted also at the Washington summit in April 1999, aiming to streamline the defence planning process and focus resources on the requirements for force projection.

Strategic missile defence was disputed, however, and formed part of a package of 'new' threats that the US sought to introduce in the Alliance but which was met with reluctance, if not outright hostility, on the part of European allies. The new threats were a mix of missile capabilities, rogue nations and international terrorism, causing the US to advocate, during the making of the Strategic Concept, a reorientation of the Alliance against these threats on a global scale. The European position was fairly clear: NATO ought to remain focused on the new grey area of crisis management which since 1989 had come to blur the distinction between war and peace on the peripheries of Europe. The end result was not comforting for proponents of allied missile defence: Europe's crisis management focus was reflected in 'The Purpose and Tasks of the Alliance', introduced early in the Strategic Concept, and NATO was moreover limited to operating in a regional zone known as the 'Euro-Atlantic area'.[19] The new threats of terrorism and missile attacks were mentioned only in a subsequent section on 'Strategic Perspectives'[20] and paragraph 56, mentioned above, appeared in the document's broad 'Guidelines for the Alliance's Forces'.

The NATO 'consensus' was adopted only after frank transatlantic exchanges in late 1998. US Secretary of State Albright failed to convince her NATO colleagues of the merits of the American agenda for a new strategic concept during a final preparatory meeting in December 1998. In the words of the British daily, *The Independent*, the US 'ran into a wall of opposition' when presenting a blueprint including 'such capabilities as an intelligence clearing house on nuclear, biological and chemical weapons and co-ordinated steps to protect the allies from attacks by such weapons'.[21] The exchanges continued beyond the April 1999 summit. In October 1999, Deputy Secretary of State Strobe Talbott sought to calm transatlantic tensions by underscoring the potential for harmonious NATO–EU cooperation and referring to the four conditions for deployment outlined by President Clinton. Talbott failed to convince, however. The *Financial Times* wrote in its editorial that the critical condition of threat assessment had already been answered in the positive by US policy-makers and that Washington, in thus failing to address the issue open-mindedly, was deluding itself and risked encouraging proliferation while also alienating Russia.[22] A month later, when Talbott provided a first extended briefing to NATO allies on US missile defence plans, he was met with strong criticism – launched first by France but seconded by Spain and then Germany, Belgium, Italy, Denmark, the Netherlands and Canada – with the new NATO allies, Poland, Hungary and the Czech Republic, making no comments. According to one observer, 'Talbott was shaken' and US officials subsequently realized that they 'had badly miscalculated the extent to which the Europeans simply would not buy the premises of the US approach'.[23]

The Clinton administration sought to take control of the situation by intensifying the dialogue with NATO allies as well as with Russia. Russia had been alienated from NATO in the context of the Kosovo war and now warned that any US decision to move beyond the ABM treaty would cause Russia to disregard disarmament treaties – in other words, to build up its nuclear forces.[24] But still, from a US perspective the relationship with Russia was of double value: it could stabilize Russian defence policy and also convince European allies that the programme could move ahead without doing damage to the international order.

In this respect the US had done itself a disservice by outlining a deployment programme that necessitated a rapid decision: it was vulnerable to international protest and hostage to technical difficulties. Time would surely ease the task of surmounting these difficulties, but time was precisely what the Clinton administration had denied itself. The year 2000 was now decisive to the US missile defence project and enhanced US vulnerability. A compressed test schedule became of essence to make an informed deployment decision by June 2000 – which was needed to begin construction in Alaska in 2001 with a view to deployment by 2005, the official deadline – but it also became the source of technical vulnerability. Moreover, the diplomatic process was compressed because the US needed to strike an

ABM deal with Russia in 2000 to begin construction in 2001, which then became a source of diplomatic vulnerability.

Clinton's policy increasingly satisfied no one. Russian President Putin in June 2000, having just met with President Clinton to agree on principles for 'strategic stability', outlined at a press conference a new radical proposal aimed at creating a united front against 'rogue nations': 'Russia proposes setting up, together with Europe and NATO, a common missile system.'[25] The Russian blueprint was in contradiction to the 'national' aspect of the Clinton blueprint and thus created conceptual confusion.

Meanwhile, within the US, critics focused on the three critical tests designed to prepare the deployment decision as well as the decision to respect the ABM engagement. With respect to the former, critics, including many scientists, derided the tests for being rigged and also alleged that three tests were insufficient to inform a decision. With respect to the latter, other critics, mostly Republicans and including presidential candidate George W. Bush, called for a completely new approach to the issue, one that simply ignored the constraints of the ABM treaty.

These attacks on the Clinton policy greatly facilitated the European status quo, which had appeared shaken in 1998–9. France and Germany were particularly active in voicing their concerns in mid-2000, as Clinton was preparing for the deployment decision, France underlining the stability flowing from a strategic policy of deterrence, Germany underscoring the need to stabilize Russia and also to turn the back on nuclear weapons rather than developing them. Britain was kinder to the Clinton policy, signalling its readiness to allow the US to upgrade its radar tracking equipment in Fylingdales, Yorkshire. The British conciliatory position could have been due to cultural affinity and the history of the special relationship; it could also be rooted in the Blair government's concern that the upcoming replacement of its Trident nuclear missile system was beholden to the US.[26] The only other European country to host radar installations critical to the American programme, Denmark, adopted a wait-and-see policy deliberately aimed at avoiding domestic controversies.

Criticized from all political directions, the Clinton policy, which just a year previously seemed to rest on a domestic consensus, was condemned to seek comfort in the last of the three missile defence tests taking place in July 2000 (the June 2000 decision having been deferred). Of the two previous tests, the first succeeded and the second failed. The third test – in the circumstances globally scrutinized – proved a failure as the booster rocket malfunctioned. In consequence, President Clinton in August deferred the decision on deployment and indeed the system's architecture to the new year and his successor who would be chosen in November 2000.

President Bush and a transatlantic division of labour?

George W. Bush was sworn in as president 20 January 2001; throughout the election campaign in 1999–2000 he had made clear his preference for a clean break with the Clinton missile defence plans in favour of a new comprehensive review of defence options. It was now evident that European sceptics could no longer rely on a moderate policy proposal, such as that promoted by President Clinton, and that allies would be forced to either align more closely with US policy or make their reservations more explicit. The reactions, again, were varied but they amounted to an increase in the level of support. Some governments came out in favour of the idea, others remained hostile to the plans but now supported a type of transatlantic division of labour according to which the US could undertake certain global responsibilities while European allies would gain greater autonomy in the regulation of European security affairs.

The Bush administration was initially not aided in its policy efforts by dramatic external developments affecting European security perceptions and it instead focused on intensifying transatlantic dialogue, coupling denser communication with a change of emphasis in that the 'national' dimension of missile defence was eliminated in favour of a new emphasis on missile defence constituting a vital component of US global involvement and leadership. Intensified dialogue was notably visible in May 2001 when a delegation headed by Deputy Defence Secretary Wolfowitz visited Brussels, Paris, London and Berlin, then in June as Secretary of Defence Rumsfeld participated in his first NATO Council meeting and Bush made his inaugural trip to Europe, where he returned just a month later.[27] Rumsfeld struck a conciliatory note, tying the plans for a missile defence to the US 'vital interest' in NATO, the desire to protect not only the US but also allies from new threats, and by minimizing the conceptual shift in strategic thinking: 'We do not intend to abandon nuclear deterrence. Rather, we see it as one layer of a broader deterrence strategy that includes several mutually reinforcing layers of deterrence.'[28]

In official documents, NATO's approach was still a balanced one, embracing continued arms control and diplomatic strategy, such as the Nuclear Non-Proliferation Treaty (NPT), and also recognizing the need to deal with threats from WMD and their means of delivery.[29] But on the sidelines of the summitry, some European governments were now voicing greater understanding of the need to think in new terms. Spain, Turkey, Poland, Italy and the Czech Republic were the most vocal allies adopting a positive view, while Britain continued its policy of supporting the plans. Bush's first stop on his European tour in June 2001 was Madrid, Spain, and Spanish Prime Minister Aznar shifted the burden of evidence to the opponents of missile defence: 'it has not been demonstrated anywhere... that the defensive initiative is something that cannot lead to greater and better security'.[30] Following a subsequent informal NATO summit, newly elected Italian Prime Minister Berlusconi spoke with President Bush at a press conference in Rome and deliberately sought to place Italy among the most enthusiastic of the allies: 'We will always be next to the United States in order to take part in this discussion, going well beyond the attitudes of certain European states, which still, today, have not, in my opinion, understood how the world has changed, and how we should start worrying about the future.'[31]

President Chirac of France remained perhaps the most outspoken sceptic of US plans, although German Chancellor Schroeder also voiced criticism. Chirac's speech to his NATO colleagues welcomed the debate on the issue but also drew attention to three principles of French policy: preserving strategic stability, involving notably the ABM treaty; combating proliferation, involving notably the NPT treaty; and supporting the overall principle of nuclear deterrence (as opposed to defence).[32]

Still, France was not insensitive to the new threats evoked by the US. First, President Chirac outlined his vision of French strategic policy at the French defence academy earlier in June 2001, using the occasion to soften the French position: he now recognized that ballistic missiles did pose a threat and moreover clearly stated the French desire to build theatre missile defences.[33] A deal on strategic defence, always the most controversial aspect, was made more likely not only by this dose of moderation but also by Chirac's view of NATO's internal evolution outlined at a press conference subsequent to the informal NATO summit, also in June 2001:

> Within the Alliance, we have highlighted the emergence of a strategic Europe (*Europe de la défense*), noting that it is a natural development and that our peers today accept it, whether in the US, Canada or Europe, with a strategic Europe representing not only a necessity for the Europeans who want to exist but also an asset for NATO and a strengthening of NATO. From this perspective things are going well.[34]

In other words, French interests would be met if European autonomy was enhanced, opening the door for a compromise where Bush's principal policy, missile defence, could move ahead without intransigent opposition from France and others. Naturally, national positions had to be finessed to allow for a compromise, but at least the potential was there.

Britain once again moved centre stage because it was the ally best placed to influence Washington's missile defence plans and it was one of the two states – the other was France – having given birth to the EU's security and defence policy. British Prime Minister Blair visited President Bush in mid-February 2001 and there reached an understanding that could provide the foundation for such a transatlantic division of labour: the US focusing predominantly on new global threats with some support from European allies, and the European allies focusing predominantly on their ability to tackle regional crises with some support from the US. While Bush in February endorsed the British vision of a European reaction force (tied to NATO but developed partly within the EU), Blair agreed to the statement that 'We need to obstruct and deter...new threats with a strategy that encompasses both offensive and defensive systems'.[35]

A transatlantic deal was not a certainty by 11 September 2001 when terrorists struck at the US and upset ongoing policy and diplomacy, but it was a likelihood. The question that emerged in the wake of the attacks was whether the deal remained valid: was the terrorist menace so compelling that the allies would join forces in a new campaign? If so, ideas for a division of labour became redundant and a new collective strategic vision had to be crafted.

A new missile deal framed by rogue nations and proliferation?

We know that the allies reached agreement in November 2002 to initiate a study on strategic missile defence and thus that the search for a common vision continued beyond September 2001. The glue of anti-terrorism proved essential in this because all allies sought to continue past policies – demonstrating how immovable policy can be – but now also sought to continue them within the new framework of anti-terrorism.

The Bush administration has not revised its ambition to build a missile defence, but it has come to focus – quite naturally – on other kinds of asymmetrical threats, very primitive in nature but overwhelming in impact, and is integrating missile defence in a comprehensive defence policy aiming to obtain 'full spectrum dominance'. Continuity and change thus coexist in US policy, implying that non-state actors and cardboard knives did not simply replace rogue nations and ballistic missiles in US threat assessments. US security doctrine instead focuses on crossroads, the intersections where radical opponents of the US, whether in control of states, sponsored by states or autonomous from states, threaten to use WMD to attack the US. This much was clear from the national security strategy published in September 2002 and which gave birth to the Bush doctrine of pre-emption. A simultaneous focus on terrorist networks and radical states was by then evident in the campaign to change the Afghan regime and uproot the al-Qaeda network, and also in the January 2002 State of the Union address in which President Bush made reference to an 'axis of evil', consisting of Iraq, Iran and North Korea.

US national security became the predominant focus of US foreign policy and policy-makers became more willing to simply ignore international constraints. In terms of missile defence, the Bush administration continued an open-ended search for a layered defence system designed to protect against both larger attacks by short- and medium-range missiles and smaller attacks by intercontinental missiles, making the decision on a system architecture dependent on technological advances and test results, but now also integrated missile defence in a more comprehensive security strategy:

- In December 2001, the US announced its withdrawal from the ABM treaty, which took effect in June 2002, thereby definitely making a break with the past framework of nuclear deterrence.
- In December 2001, the US Nuclear Posture Review (NPR) presented a new plan for a nuclear 'triad' consisting of offensive weapons (including 'a more effective earth penetrator' – a small nuclear weapon), defensive systems (including missile defence) and a modernized production infrastructure.[36]
- In May 2002, the US and Russia signed a new nuclear arms reduction agreement by which the two countries pledged to reduce their offensive nuclear arsenals to about 2000 weapons each by 2012. The agreement placed no restrictions on missile defence, and in contrast to the previous Strategic Arms Reductions Talks (START) provided freedom of choice in respect to the composition of the nuclear arsenal and the decision to store some nuclear arms (which then do not count in the offensive arsenal comprised by the agreement).
- In December 2002, the National Plan to Combat Weapons of Mass Destruction integrated missile defence as one component of 'counter-proliferation' – the first of three policy pillars, the two others being non-proliferation and consequence management.[37]

The comprehensive approach to security involved not only the integration of weapons programmes into political documents, but also the gradual support for international diplomatic solutions to pressing developments that could result in the type of threat invoked by missile defence proponents. Such international diplomatic solutions were at the heart of European policy and constituted the necessary dimension of continuity that made it possible for several European allies to move closer to the US missile defence option. Two critical cases illuminate the point.

In late 2002, North Korea announced its withdrawal from the non-proliferation treaty (NPT) and thus threatened to begin anew its production of weapons-grade plutonium and enriched uranium (it had a large stock of spent fuel rods for the purpose which until December 2002 was sealed off by the International Atomic Energy Agency [IAEA]), and in the spring of 2003 it restarted a nuclear reactor that had been shut down as part of the 1994 agreement that also involved the IAEA's control of the fuel rods. Months later it declared itself to be in possession of enough plutonium to build six nuclear bombs. Two aspects are noteworthy in this case. First, European governments sought close cooperation with the US when in early

2003 they decided to plan for a diplomatic mission to Pyongyang (mandating the Greek EU presidency to prepare the mission). According to one EU diplomat, 'There have been no lines crossed between Brussels and Washington. No competing in the region. No breakdown in communication between both sides of the Atlantic.'[38] This was remarkable in light of a similar EU mission of May 2001 which was poorly coordinated with the US and therefore perceived as constituting an implicit criticism of US policy.[39] Second, the US supported the EU initiative, breaking with the Bush administration's reluctance to negotiate with Pyongyang and its refusal to pursue the agreement that President Clinton was very close to reaching with the North Korean leadership in December 2000.[40]

In relation to another of the countries included in Bush's 'axis of evil' speech, Iran, which in the course of 2003 was revealed to maintain a nuclear development programme that in all likelihood had a military dimension (the IAEA concluded in November 2003 that Iran had a programme for developing weapons-grade plutonium), the US and European allies were in agreement that Iran's nuclear programme should be arrested, although they managed to fall out over the means. A crisis related to the demonstrable readiness of the US and Israel to strike pre-emptively against WMD programmes was defused in October 2003 when a troika of ministers from Britain, France and Germany secured Iran's consent to a deal that would suspend the Iranian enrichment programme and allow unhindered inspections, a deal that was initially criticized in the US for whitewashing Iran's WMD programme and for failing to include a trigger mechanism in case of Iranian non-compliance.[41] Transatlantic differences have narrowed, however, owing to Iran's subsequent refusal to abide by the agreement and its announcement, following its censuring by the IAEA in June 2004, that it would resume its enrichment programmes.[42]

Two contrasting lessons can be learned from these 'rogue cases'. On the one hand, European allies are becoming more serious about proliferation and the danger that failed states and regional conflict could produce a missile threat. They are also in agreement with the US that missile defence needs to be integrated in a comprehensive security approach, forming part of a package that includes not only a military capability to deter (such as offensive weapons) but also a wide range of political-economic instruments.

On the other hand, within this package, European governments tend to place more emphasis on the political-economic instruments and downgrade the role of weapons. European leaders thus seem more convinced that they can do business with moderates within Iran, and refused to tie the IAEA censoring in mid-2004 to a timeline and the imposition of sanctions, which the US favoured. Some European leaders also tend to be more critical of the claim that some state leaders cannot be trusted to interact rationally (and thus be deterred), which accounts for the transatlantic divide that opened up in late 2002 on the issue of

Iraq, its WMD programme and political intentions, and the divide is certainly not being healed by the failure of American and British forces to find WMD in Iraq. It would seem that several European governments, such as the French and the German, perceive 11 September to have invoked an era in which ballistic missile threats, while important, *recede* in importance compared to failed states and terrorism, whereas the Bush administration perceives 11 September to have revealed a type of threat that is *as serious as* that of ballistic missiles.

As in the case of the Iraq war, European allies are split. They are able to unite when it comes to endorsing 'soft' security strategies that emphasize 'preventive engagement' and 'effective multilateralism' – euphemisms for the established policy of dialogue with international renegades.[43] In specific contexts dealing with nations that may pose a missile threat, they tend to split.

Britain and Russia are the two large European countries that have aligned most closely with the US agenda.[44] President Putin was the first to call President Bush in the wake of the 11 September attacks, and he has since committed Russia to a pattern of cooperation that involves a permanent US military presence in the former Soviet Union (i.e. Central Asia), a course of action that no doubt sparks hostility in wide sections of the Moscow security complex. Putin also did not protest when the US terminated the ABM issue by simply withdrawing from the treaty, and Putin accepted the new, looser version of a strategic arms agreement sought by Bush (from May 2002) and also accepted the second and geographically extensive round of NATO enlargement announced in November 2002. Putin, in short, is staking a significant amount of political credibility on the fight against international terrorism and the likelihood that the US will reward Russia one way or another for its participation.[45]

British Prime Minister Blair is supportive of the US missile defence plans, as we have seen, and has also supported Bush through the toughest phases of the Iraq campaign. Moreover, British policy deviates interestingly from the French one in respect to Russia, a country that both France and Britain are eager to anchor in the West. Whereas France seeks primarily to tighten relations between the EU and Russia – for many good economic reasons – while also soothing Russian pretensions with the established French discourse on multipolarity, Britain has sought to bring Russia on board NATO in the fight against terrorism. In fact, in the immediate wake of 11 September, Blair may have offered Putin, without consulting in advance with Bush or other NATO leaders, to 'arrange Russia's passage into the NATO sanctum' – because Western and Russian strategic interests were becoming alike.[46] The end result did not quite match the original idea: Russia gained access to a new NATO–Russia council established in May 2002 (which in any case would have been necessitated by the upcoming enlargement decision because, in order to enlarge, NATO needs to bolster cooperation with Russia) and whose purpose is to examine, among other issues, international terror, proliferation and also missile defence.

Britain is seeking to develop its engagement in Europe as well as across the Atlantic, which is why Britain sometimes finds itself in opposition to the US, as was the case in Iran in late 2003. Still, developments in British and Russian policy indicate that the past division of labour is off the table. The EU may still do regional crisis management while the US – perhaps through NATO – focuses on new threats on a global scale, but the division between the two has been blurred significantly by the common interest in fighting international terrorism, which, in some cases, may involve state-sponsored terror and missile threats. At the very least, this will demand of NATO a reform to broaden its gaze beyond the confines of the Euro-Atlantic region and consider ways of involving Russia on meaningful terms. NATO began this type of transformation at the Prague summit in November 2002, where, as stated at the beginning of this chapter, the NATO allies also initiated a study of missile defence. NATO may succeed in crafting a united stand on the issue, but success is contingent on several big questions: Will Putin manage to sustain Russia's constructive engagement with the West? Will France gain enough on the European front to positively engage in NATO's transformation? And will technological developments in the US allow the Bush administration (or its successor) to craft a missile defence blueprint that is both feasible and affordable? In lieu of answers we turn to the conclusion.

Conclusion: the future is rogue

European reactions to US missile defence plans are varied, but through the 1990s to the present a pattern emerges whereby European reluctance as well as hedging gradually give way to a more positive view. It is not as if the positive attitude is of immediate comfort to Washington: it amounts to an agreement to study the issue and more generally to consider the importance of missile threats, but it is predicated on political divisions that tend to erupt in concrete instances.

As a rule of thumb, Europeans have been swayed on the subject whenever regional crises involving a rogue nation possessing either missiles or WMD coincide with specific US plans for missile defence. In 1998–9 this was happening (with European governments preparing to make concessions by announcing their firm opposition to any concessions), but the process unravelled along with the Clinton missile defence policy. In early 2001, clear signals from the Bush administration, unaccompanied by new regional developments, resulted in a tentative agreement to disagree: each side of the Atlantic would pursue their area of specialization and subsume it under the heading of renewed partnership.[47] The terrorist attacks of September

2001 brought to light external security developments, but they were not directly related to missile developments and therefore raised new questions in relation to US missile defence plans. The US responded by incorporating missile defence in a comprehensive security strategy, European governments by accepting the need to focus on new threats but also continuing their focus on diplomatic engagement.

For NATO to succeed in tying the two sides of the Atlantic together on the issue of missile defence, short- and long-range measures must be included in a policy package. European governments are as ever willing to address long-range measures – promising for instance to ameliorate Pakistan's social and economic conditions in the hope of inciting political reforms. The US tends to worry about short-term measures to counter deceit – providing insurance against rogue leaders' intentions.

Observers may suggest that nuclear proliferation will be a source of transatlantic crisis,[48] but on the issue of missile defence it seems that a consensus is emerging and that the issue will gain a more prominent place on NATO's agenda. This estimate is based on several observations. First, the European allies at Prague made a principled commitment to missile defence that they cannot turn away from, even if they wanted to. Moreover, a number of developments are making it easier for European governments to continue their careful embrace of missile defence: the debate has been going on in Europe for so long that by now it is easier to make a rhetorical switch from 'whether' to 'when'; the comprehensive approach of the US to WMD issues allows them to address missile defence in a 'balanced' context; they can justify a more positive attitude with reference to a window of opportunity for gaining influence on an allied command and control arrangement;[49] and they can also justify such an attitude on grounds of economic pragmatism.[50]

All this is not to say that transatlantic missile defence cooperation is a foregone conclusion. Disputes over Iraq between the US and Britain on the one hand and France on the other could be damaging in other contexts: for instance, France was likely uninformed of the US–British diplomacy that led Libya to renounce its WMD programmes in December 2003.[51] Moreover, the most critical issue will be rogue nations, whose behaviour tends to be unpredictable and whose domestic politics are obscure to outsiders. Two critical rogue nations are Iran and North Korea; another is Pakistan, which was recently designated a 'major non-NATO ally' by the US though it is now identified as a decisive hub in a nuclear proliferation network.[52] Still, for the reasons enumerated above and discussed in the analysis, it seems likely that the NATO allies will deepen their agreement to develop a missile defence capability and also strengthen the coordination of their policies vis-à-vis rogue nations.

Notes

1 NATO, *Prague Summit Declaration*, 21 November 2002, paragraph 3.
2 Ibid., points e and g of paragraph 4.
3 NATO, *NATO Missile Defence Feasibility Study: Transatlantic Industry Study Team Selected*, NATO press release 109, 26 September 2003.
4 US Department of State, 'NATO Defence Ministers Invigorated by Discussion in Colorado', *International Information Program*, 9 October 2003. Online. Available HTTP: ⟨http://usinfo.state.gov/topical/pol/nato/03100919.htm⟩.
5 NATO, *Istanbul Summit Declaration*, 28 June 2004, paragraph 12.
6 See General Sir Hugh Beach, 'Theatre Missile Defence: Deployment Prospects and Impact on Europe', *ISIS Briefing on Ballistic Missile Defence*, no. 2, September 2000.
7 The agreement was reached between Presidents Clinton and Yeltsin in June 1997 when the two state leaders were also wrestling with NATO enlargement and strategic arms reduction talks (cf. START).
8 American Foreign Policy Council, 'A Push Toward Europe from the West', *Missile Defence Briefing Report*, no. 64, 31 July 2002. Online. Available HTTP: ⟨http://www.afpc.org/mdbr/mdbr64.htm⟩.
9 ISN Security Watch, *US Eyes 'Star Wars' Interceptor Base in Central Europe*, news list distributed 14 July 2004.
10 NATO, *Istanbul Summit Declaration*, 28 June 2004, last bullet of paragraph 19.
11 Congress of the United States of America, National Missile Defense Act of 1999. Online. Available HTTP: ⟨http://www.cdi.org/hotspots/missileDefense/act.html⟩.
12 The two previous occasions were in the late 1960s and early 1990s, respectively, when the Nixon administration pursued the Sentinel/Safeguard system and the Bush administration downscaled the Strategic Defence Initiative to a defence capability, albeit global, against limited strikes (known as GPALS). The consensus was fragile in both cases, as it turned out to be also in the late 1990s. See Bradley Graham, *Hit to Kill: The New Battle over Shielding America from Missile Attack*, New York: Public Affairs, 2001.
13 Ibid., pp. 32–34.
14 *Report of the Commission to Assess the Ballistic Missile Threat to the United States*, 15 July 1998, reprint available online at HTTP: ⟨http://www.fas.org/irp/threat/missile/rumsfeld⟩; the citations are from the summary.
15 The report noted that 'Iran is making very rapid progress in developing the Shahab 3 MRBM' and that 'This missile may be flight tested at any time and deployed soon thereafter'. The July test was generally a failure, however, as the Shahab 3 exploded shortly after its launch.
16 John Isaacs, 'Rumbles from Rumsfeld', *Bulletin of the Atomic Scientists*, 55(4), September–October 1998.
17 Public Broadcasting Service Frontline, *Missile Wars*. Online. Available HTTP: ⟨http://www.pbs.org/wgbh/pages/frontline/shows/missile⟩.
18 NATO, *The Alliance's Strategic Concept*, 23–24 April 1999, NAC-S(99)65, paragraph 56. Online. Available HTTP: ⟨http://www.nato.int/docu/pr/1999/p99-065e.htm⟩.
19 European adherents of tying NATO to a territorial defence role and thus opponents of NATO's global involvement were not all happy with this definition of NATO's regional limits because the boundaries of the 'Euro-Atlantic' area are vague and can be stretched by political circumstance. Still, the focus was less than what US policy-makers had hoped for.

20 NATO, *The Alliance's Strategic Concept*, 23–24 April 1999, NAC-S(99)65, paragraphs 22–24.
21 'Germany and the US Split on NATO', *The Independent*, 9 December 1998; also *New York Times*, 9 December 1998.
22 'License for a Unilateral America', *Financial Times*, 29 October 1999.
23 Graham, op. cit., pp. 154–156.
24 'No Headway on ABM Adaptation Nor Will There Be', *Interfax*, 22 October 1999.
25 'Putin Urges US to Build Joint Nuclear Shield', *Financial Times*, 3 June 2000; 'Putin Proposes Joining West in Missile System', ibid., 6 June 2000.
26 Philip Stevens, 'An Umbrella Against the World: The US Believes it Can Shut out the Risk of Nuclear Attack with Technology', *Financial Times*, 14 April 2000.
27 On the new Bush approach, see Thérèse Delpech, 'A New Transatlantic Deal on Missile Defense After the Terrorist Attacks?', revised version of paper presented at the Pugwash Workshop on Nuclear Stability and Missile Defence in September 2001. Online. Available HTTP: ⟨http://www.puwash.org/september11/delpech.htm⟩.
28 Donald Rumsfeld, *Prepared Remarks by Secretary of Defense Donald H. Rumsfeld, Brussels, Belgium*, 7 June 2001. Online. Available HTTP: ⟨http://www.defenselink.mil/speeches/2001/s20010607-secdef.html⟩.
29 See in particular NATO, *Final Communiqué*, M-NAC-1(2001)77, 29 May 2001, paragraphs 76–79.
30 Cited in David J. Smith, 'Bush European Success Calls for NATO Missile Defence Plan', *Inside Missile Defence*, 7(14), July 2001.
31 Transcript: *Bush, Italian PM Berlusconi Press Conference*, Rome, 23 July 2001, Online. Available HTTP: ⟨http://www.nato.int/usa/president/s20010723a.html⟩.
32 Jacques Chirac, *Intervention de Monsieur Jacques Chirac, Président de la République, lors de la Réunion Spéciale du Conseil de l'Atlantique Nord*, Brussels, 13 June 2001.
33 Jacques Chirac, *Discour devant l'Institut des Hautes Etudes de Défense Nationale*, 8 June 2001. See also 'France Softens its Stance on US Plan for Missile Shield', *Financial Times*, 9–10 June 2001, p. 3.
34 Jacques Chirac, *Conférence de Presse de Monsieur Jacques Chirac, Président de la République, à l'Issue de la Réunion Spéciale du Conseil de l'Atlantique Nord*, Brussels, 13 June 2001.
35 'Bush and Blair in Tune on Vital Defence Issues', *Financial Times*, 24 February 2001.
36 Excerpts from the NPR were made public and submitted to Congress in December 2001. See *Nuclear Posture Review Report* Online. Available HTTP: ⟨http://www.globalsecurity.org/wmd/library/policy/dod/npr.htm⟩. David Yost notes that the NPR's focus on goals other than merely deterrence has been 'disquieting to Allies': 'The US Nuclear Posture Review and the NATO Allies', *International Affairs*, 80(4), July 2004, p. 707.
37 *National Strategy to Combat Weapons of Mass Destruction*, December 2002. Online. Available HTTP: ⟨http://www.state.gov/documents/organization/16092.pdf⟩.
38 'Europeans and US Work Together for N. Korea solution', *Financial Times*, 5 February 2003.
39 For instance, Henry Kissinger criticizes the policy of the EU and in particular that of Sweden, the driving force behind the 2001 EU mission. 'They should ponder', Kissinger writes, 'whether, in taking this course, they are not bringing

about the exact opposite of their proclaimed intentions.' Henry Kissinger, *Does America Need a Foreign Policy? Toward a Diplomacy for the 21st Century*, New York: Simon & Schuster, 2001, p. 130.

40 For an account and general criticism of the Bush policy, see John Newhouse, *Imperial America: The Bush Assault on the World Order*, New York: Alfred Knopf, 2003, ch. 4. The European mission to North Korea never took place because Pyongyang opted for a three-way summit with Washington and Beijing, and the EU has since faded from the diplomatic scene, as this summit gave birth to six-nation negotiations, involving the US, North Korea, China, South Korea, Russia and Japan, which continued with modest success in 2004 and 2005.

41 'See No Evil', *Washington Post* editorial, 24 November 2003; 'New Discord Rises Between US and Allies', *Los Angeles Times*, 19 November 2003.

42 Iran signed but has yet to ratify the Additional Protocol to the NPT that allows for unhindered inspection. Therefore, in June 2004, the IAEA 'deplores' that 'Iran's cooperation has not been as full, timely and proactive as it should have been'. *Implementation of the NPT Safeguards Agreement in the Islamic Republic of Iran*, resolution adopted by the IAEA Board of Governors, 18 June 2004. Online. Available HTTP: ⟨http://www.iaea.or.at/Publications/Documents/Board/2004/gov2004-49.pdf⟩. The IAEA reported in August 2004 that some of the enriched uranium found in Iran may have originated in Pakistan, raising new questions about whether Iran accidentally imported the uranium in black-market equipment and whether Iran in fact does maintain a nuclear weapons programme.

43 See the European Union Security Strategy, *A Secure Europe in a Better World*, 12 December 2003. Online. Available HTTP: ⟨http://ue.eu.int/pressdata/EN/reports/78367.pdf⟩.

44 Poland, as the introduction noted, is another staunch supporter of US policy. Denmark, host to a critical US radar site in Greenland, likewise supports US missile defence policy. In August 2004 the Danish and American authorities entered an agreement allowing for the radar site's upgrade.

45 This may be part of the reason why Bush's national security advisor, Condoleezza Rice, in the wake of the Iraq war promised to deal with the principal critics on different terms: forgiving Russia, ignoring Germany, and punishing France.

46 Newhouse, op. cit., p. 145.

47 Charles Ball concludes that with the election of George W. Bush, the European allies viewed deployment as inevitable and thus opted for a more positive attitude. This analysis agrees only partly: it took September 2001 to produce the positive attitude and up to then the Europeans were more focused on getting 'something for something' – indicating their continued scepticism towards the idea of missile defence. See Charles Ball, 'The Allies', in James J. Wirtz and Jeffrey A. Larsen (eds), *Rockets' Red Glare: Missile Defences and the Future of World Politics*, Boulder, Colo.: Westview Press, 2001, pp. 259–279.

48 Fareed Zakaria, 'Iran: The Next Crisis', *Washington Post*, 10 August 2004.

49 David Yost underscores the importance of this time pressure inherent in the command and control issue. Yost, op. cit., pp. 720–721.

50 European procurement on theatre missile defence capabilities resulted in a multi-billion Euro contract in late 2003, and some of the beneficiaries of this contract, an international missile defence systems group composed of European companies, in mid-2004 signed a memorandum of understanding with the US defence companies Lockheed Martin and Northrop Grumman with the intention of jointly pursuing 'global ballistic-missile defence opportunities'. See 'OCCAR

Awards 3 Billion Euro Missile Contract', 13 November 2003. Online. Available HTTP: ⟨http://www.defence-aerospace.com/produit/28701_us.html⟩, and *Reuters*, 'Northrop, EADS to Seek Joint-Missile Defense Work', 22 July 2004. Online. Available HTTP: ⟨http://www.reuters.com/newsArticle.jhtml?type=reutersEdge&storyID=5748336⟩.

51 'How the Deal Was Done', *Sunday Telegraph*, 21 December 2003.

52 'Rogue Nuclear Projects: Tangle of Global Clues Has Pakistan at Centre', *International Herald Tribune*, 5 January 2004, and 'Trails of Rogue Nuclear Projects Lead to Pakistan', ibid., 6 January 2004; 'Nuclear Expert "Admits Selling Secrets"', *The Guardian*, 3 February 2004.

6

CHINA'S RESPONSE TO THE US MISSILE DEFENCE PROGRAMME

National options and regional implications

Peter Dyvad

Introduction

This chapter takes its point of departure in the assumption that the security policy dynamic in East and South Asia depends to a great extent on the interaction between the United States and China. Official American declarations have explicitly emphasized the fact that the plans to develop and deploy a missile defence (MD) are based exclusively on strategic considerations of protecting the US and its allies against attack from 'rogue states'.[1] MD is also intended to protect the US against accidental and unauthorized missile launches from states that could be thought to have ballistic missiles directed against it, i.e. primarily Russia, China and rogue states.[2] Despite the American declarations, however, the fact is that members of the George W. Bush administration have pointed to China as the 'strategic rival' of the US over the long term and thereby indicated the presence of great power interests behind the incentives for MD.

Beijing's view that MD is directed towards China is based on the fact that the system has an inherent capacity to be able to eliminate China's minimalist nuclear deterrence. The MD architecture is clearly oriented primarily towards missiles launched against the US from East Asia and the system is expected to obtain a capability that counteracts China's small number of ICBMs.[3] At the outset, X-band radar[4] and interceptors will be deployed first and foremost on Shemya Island in Alaska, a geographically well-located outpost to detect and track incoming ballistic missiles launched from Russian Siberia, North Korea and China.[5] By optimizing MD to engage ballistic missiles from North Korea, the US would at the same time also have optimized the system to defend against China's ICBMs. Moreover, the East Asian focused architecture of MD gives Beijing a reinforced impression that ballistic missiles from East Asia would have only a poor chance of penetrating MD.

As a growing great power, China seeks parity with the US. In this connection, it cannot be ruled out that the security policy relations between China and the US could develop into a conflictual direction, where military confrontation, including the threatened use of nuclear weapons, could take place. At a minimum, the nuclear balance of power will influence the two actors' options and negotiating positions, and in this connection Chinese parity with the US will entail the maintenance of a credible nuclear deterrence. The aim of this chapter is to examine and explain, relying on 'balance of threat' theory, how the great power aspect has decisive significance for China's response to MD and to demonstrate that this response will also have regional implications in East and South Asia.[6]

The chapter begins by analysing China's perception of aggressive American intentions. It then discusses China's national options to balance MD. Finally, certain regional implications of China's counterbalancing strategy will be identified. The chapter ends with a brief conclusion.

Aggressive American intentions

The political context

The American missile defence plans arouse great anxiety in China, in that there is a fear that MD will undermine China's nuclear deterrence vis-à-vis the US. Beijing's attitude towards MD is reflected in the official Chinese rhetoric, which indicates an uncompromising opposition towards the system.[7] Beijing has characterized MD as unilateral nuclear expansion, accusing the US of pursuing a strategy based on pre-emptive application of nuclear weapons. Under such circumstances, MD would serve as a force multiplier for American offensive strategic forces.

Beijing's anxieties about a subversion of China's nuclear deterrence must be seen in light of Sino-American relations in general, and in the context of the American engagement and intentions in East Asia in particular. The Chinese leaders are easily capable of identifying several scenarios where the relationship between China and the US can evolve in a conflictual direction. The US has been one of the most prominent critics of what Beijing considers to be internal Chinese affairs, these being the human rights situation in China, the fight against 'terrorists',[8] and especially China's view of Taiwan as a 'renegade province'.

China can also conclude that several of the bilateral agreements in the San Francisco system, created during the Cold War to contain communist expansion, have been maintained even after the threat from the Soviet Union has disappeared, and that formalized military cooperation – especially with Taiwan, Japan, South Korea and the Philippines – forms the basis for continued American engagement and power projection in East Asia.[9] China regards the US as the dominant power in East Asia and thereby also the greatest obstacle to China's possibilities of pursuing its security interests in the region; these interests include Chinese territorial claims over Taiwan, the South China Sea and the Senkadu (Diaoyu) Islands in the East China Sea.[10] China does not exert de facto control over these areas but has asserted its right to pursue these territorial demands with the application of military force.[11] The US does not counter every Chinese move – for instance, it did not react when China occupied the South Vietnamese Paracel (Xisha) Islands in 1974, or the Mischief Reef in 1995, which was also claimed by the Philippines – but is likely to counter significant Chinese initiatives.

This was made especially evident to Beijing in July–August of 1995 and again in March 1996, when the US deployed two aircraft carrier groups to the Strait of Taiwan as a response to the Chinese demonstration of power intended to intimidate Taiwan, which was holding parliamentary and presidential elections.[12] The Chinese demonstration of strength included both the deployment of ballistic missiles to Fujian province, from which targets in Taiwan could be reached, and the conducting of military exercises which included the test firing of four DF-15 missiles in immediate proximity to Taiwan.[13] Ballistic missiles play a key role in the military threat with which China confronts Taiwan, and China is estimated to possess about 400 missiles based permanently in Fujian province. In case of a crisis, it is presumed that Beijing could rapidly increase the number of these missiles.

During an official visit to China in February 2002, President Bush also emphasized that American policy on the Taiwan question remains unchanged: the US will not accept Chinese provocations and will continue to support the Taiwan Relations Act. In addition, advisers to President Bush have in the past strongly supported Taiwan in opposition to mainland China, and these advisers have been influential during President Bush's first presidential term.[14] The politics of Taiwan in Washington is thus a cause of concern for Beijing, which considers the reunification of China and Taiwan as the most important Chinese security issue.[15]

All evidence indicates that the Strait of Taiwan continues to remain one of the most dangerous flashpoints in East Asia. Even though a large military confrontation has been avoided, many factors of instability continue to keep alive the risk of the escalation of conflict, including especially China's military build-up, Beijing's assertion of the right to use force and Taiwan's growing national identity and democracy. Beijing will not accept the Taiwan conflict becoming the object of any kind of external interference, an attitude that has contributed to the fact that the US engagement in the conflict, extending back to 1949, has been the most central problem for Sino-American relations.[16]

Three principal factors have fed Chinese scepticism concerning American intentions: (1) the US is spreading TMD[17] systems to China's neighbours; (2) the US characterizes China as a 'revisionist state'; and (3) the US has historically attempted to neutralize China's nuclear deterrence.

The United States as a TMD proliferator

China's rise to great power status and the continuing American engagement in the region raise the question of how these two powers can influence the policies of lesser powers and thereby maintain a competitive edge. Unsurprisingly, China worries about the US ability to deploy TMD systems around China via its alliances with Taiwan, Japan and South Korea. Such TMD systems would limit China's possibilities of threatening or using military force on its neighbouring states, and Beijing has strongly opposed every type of development and deployment of missile defences in East Asia. This was made clear in a joint declaration by China's President Jiang Zemin and Russia's President Putin, which stated that the incorporation of Taiwan into any foreign missile defence system is unacceptable and would seriously harm stability in the region.[18]

If regional TMD systems in East Asia become integrated into MD via, for example, common early warning satellites and communication systems, it would not be possible to separate the two systems' architecture and mission. Beijing has therefore emphasized that the architecture and mission of MD is the most important issue in connection with the proliferation of TMD by the US in East Asia. If TMD comes to determine the forward part of MD in the region, its influence on regional security and stability in East Asia will become as equally negative as the system's undermining of the global strategic balance and stability.[19]

The statements by Secretary of Defense Rumsfeld that one cannot unequivocally distinguish between 'national' and 'theatre'[20] probably feed Beijing's fear that the Bush administration's ambitious MD, which possibly includes land-, sea-, air- and space-based elements, not only has the purpose of undermining China's nuclear deterrence in relation to the US, but is also intended to limit China's offensive capability towards its neighbours. In cases of a future Sino-American conflict over Taiwan, China could thereby see itself subjected to the same nuclear blackmail situation as occurred at the end of the Korean War and during the crises in the Taiwan Strait over the islands of Quemoy and Matsu in the 1950s.

Beijing's view of American cooperation with Taiwan on TMD is thus dominated by a political dimension that overshadows even the military strategic significance of the system, because this cooperation is viewed as an expression of the engagement of the US in the defence of Taiwan. A missile defence of Taiwan would also entail a very close military operative cooperation between the US and Taiwan, something that Beijing would view as a de facto revival of the US–Taiwan security alliance, which the US agreed to terminate in 1979.

China as a revisionist state

Washington has repeatedly maintained that the deployment of MD is exclusively motivated by the threat from rogue states and by the risk of accidental or unauthorized missile launchings from Russia or China. Beijing's reaction to this policy declaration, however, indicates that they are not absolutely convinced that MD is not directed primarily towards China from a strategic point of view. The American threat perception centring on rogue states is viewed by Beijing as exaggerated, and Beijing does not trust the American view that these states' missile capacities constitute a genuine threat to the US. In the Chinese view, deployment of MD is not the correct way to counter proliferation of ballistic missile technology and weapons of mass destruction. Instead, this threat should be balanced by the use of political and diplomatic means.[21] Similarly, the American argument concerning the risk of accidental or unauthorized Chinese missile launchings appears to lack credibility seen from the Chinese perspective, as the Chinese ICBMs have a very low level of launch readiness.[22]

China's view of the real rationale behind MD is supported by statements in the domestic American political debate that have referred to China as a revisionist state and as the most important opponent to the US in the power struggle over influence in Asia. National Security Advisor Condoleezza Rice stated prior to the presidential elections of 2000 that China, as a revisionist opposed to the status quo, is a rival to the US in the Asian–Pacific region.[23]

Several events have also contributed to creating a widespread Chinese view that the US, in pursuing its national interests, acts unilaterally, without regard to either the UN Charter or to Chinese sovereignty. This view has been reinforced especially in connection with NATO's bombardment of Kosovo in 1999, the American bombing of the Chinese embassy in Belgrade, the emergency landing of an American EP-3 reconnaissance plane in China in 2001, the continuing American political and military support of Taiwan, and the American-led intervention in Iraq.

The historical context

The manifest Chinese suspicion towards both the American intentions and the official American argument for MD can also be understood in their historical context. Beijing's view that MD is intended primarily to undermine China's nuclear deterrence towards the US, and that China is unofficially viewed by the US as a competitor, is supported by the fact that the US had previously had a missile defence 'officially aimed' at China. In 1967, when the Johnson administration acknowledged its plans for the deployment of the Sentinel system, the US realized that it was not advisable to deploy a missile defence that was designed to engage Soviet missiles. The Sentinel system should be 'non-threatening' to the Soviet Union and was therefore instead officially designed to protect the US against a so-called 'Nth country threat', which included attacks by a small and uncomplicated number of ICBMs as well as accidental missile launchings from the Soviet Union, China or a third party.

In this way, Secretary of Defense Robert McNamara somewhat unconvincingly diverted the Sentinel system from being anti-Soviet to anti-Chinese.[24] Furthermore, this 'turn-around' was very appropriate because the Sentinel system had only the capability to combat a limited threat with a small number of uncomplicated ICBMs, which China began developing following its first nuclear test in 1964. The US expected that China would already have operational ICBMs by 1970; however, the American estimate was erroneous and the first Chinese ICBMs were not deployed until 1980.

When the Nixon administration came to power in 1969, the anti-Chinese rhetoric disappeared with the deployment of the Safeguard system. Thus, Beijing experienced within a relatively short period of time two radical reversals in Washington's rhetoric and a mistaken perception of China's military technological capacity. Hence, history may teach Beijing that political intentions change but also that military capabilities represent durable political facts. A durable Chinese strategy will therefore focus as much on these facts – and the need to counter them – as on the current political mood in Washington.

China's response: national options

The nuclear balance of power between China and the US is not based on parity or strategic arms control agreements. In both quantitative as well as qualitative terms, China is markedly inferior to the US in all strategic capabilities. The American offensive strategic forces stand in profound disproportion to a Chinese strategic 'triad'[25] consisting of only 20 ageing silo-based CSS-4 ICBMs that can reach American territory. China does not possess intercontinental bombers, and the country's only operational SSBN[26] – which operates exclusively in Chinese territorial waters with its 12 medium-range SLBMs[27] – poses no genuine threat to American territory and is assessed as having a very poor survivability vis-à-vis the superior ASW[28] capacity of the US. Whereas the US, in comparison, possesses

1,238 delivery platforms and 5,949 warheads, China has – not even on a quantitative basis – no counter-force[29] capability compared to the US.

Since China deployed its first ICBMs in 1980, nuclear deterrence towards the US has therefore been based more on quantitative uncertainty about China's nuclear weapons arsenal than its actual size. The approximately 20 silo-based ICBMs that the US intelligence community estimates that China possesses have a very poor chance of surviving a pre-emptive American nuclear strike. However, China has never confirmed or denied the American estimates regarding its nuclear arsenal, and it will therefore be difficult for the US to rule out a certain margin of error. China's nuclear strategy is thus based on a counter-value doctrine that aims to deter the US with the prospect that China, following a pre-emptive American nuclear strike, will be able to retaliate by launching a small number of unidentified or surviving ICBMs against major American cities.[30] Hence, China does not possess a genuine survivable nuclear deterrent in the same way as the concept is viewed by the major nuclear powers.

Beijing has always declared very consistently and openly that the Chinese nuclear weapons strategy is exclusively defensive in nature. Beijing has repeatedly renounced any first use of nuclear weapons, claiming that China's nuclear forces were developed solely with the intent of defending the country's national security interests against the possibility of nuclear blackmail.[31] This factor must be seen primarily against the background of the Chinese experiences during both the Korean War (1950–3) and in connection with the crises in the Strait of Taiwan in the 1950s, when China was subject to unequivocal American nuclear blackmail.[32] The maintenance of a survivable nuclear deterrence has therefore had highest priority in Chinese defence planning ever since the latter half of the 1950s, and analysts assess that this is also the case today.[33]

The pre-existing Chinese anxieties about the survivability of Chinese retaliatory forces have been further reinforced by the prospect that the small number of surviving Chinese missiles must then penetrate MD.[34] In conjunction with the Chinese anxieties, it must be added that China's first-generation ICBMs, with their slow-burning boosters, single warheads and total lack of countermeasures, will be very vulnerable to even a 'limited' MD. In addition, China sees itself confronted with an American missile defence that will grow both quantitatively and qualitatively and challenge China's strategic modernization programme. As long as China continues to modernize its nuclear strategic forces at what the US intelligence community has assessed as the greatest possible speed, China will probably possess 75–100 ICBMs that can reach targets in the US by 2015.[35] Against this background, the original plan to deploy 100 interceptors by 2010 and possibly up to 250 interceptors by 2015 would entail that MD would continually pose a serious threat to China's retaliation capability.

In this context, the Bush administration's very ambitious plans for the future capability of MD appear even more threatening to Beijing. Similarly, changes in the American rhetoric regarding both the architecture and timetable for deployment of MD contribute to Chinese insecurity.[36] If the Bush administration's MD is necessary in order to defend the US against a small number of missiles from rogue states, the American security policy motives must in the best case appear rather unclear to Beijing. As long as the Chinese analyses indicate that the MD architecture is significantly more ambitious than necessary in order to defend the US against rogue states, it would be logical for Beijing to conclude that the US clearly has aggressive intentions towards China.[37]

As a consequence of its anxieties about the survivability of its strategic retaliatory forces, China, since the mid-1980s, has initiated development programmes intended to deploy a modern, mobile and more survivable strategic deterrence vis-à-vis both the US and Russia.[38] The ongoing development of China's strategic arsenal is aimed at the acquisition of a deterrence strategy, which is not dependent on uncertainties about a strategic opponent's estimate of the size of the Chinese nuclear force. This implies that, regardless of how precisely the US will be able to determine China's total number of nuclear weapons, at least a small number of the Chinese ICBMs or SLBMs must be able to survive a pre-emptive American nuclear strike. This change in China's nuclear strategy thus has the aim of avoiding every form of uncertainty regarding China's ability to retaliate to a pre-emptive counter-force strike.[39] China is therefore developing three new, mobile, solid-propellant strategic missiles: the DF-31, a land mobile ICBM with a range of 8,000 km and which is expected to be deployed before 2005; the DF-41, a long-range version of the DF-31 with a range of 12,000 km; and the JL-2, a new SLBM system with a range of about 8,000 km and which is expected to be deployed by 2010.[40]

China is expected to multiply its strategic forces several times over the next 15 years, just as the majority of the strategic ballistic missiles will probably have become mobile. The intelligence community estimates that by 2015, China will have 75–100 warheads deployed in ICBMs able to reach the continental US. Other analyses estimate that China will have up to 102–190 warheads deployed in ICBMs by 2015, in that the total number will be dependent on the application of MRVs[41] or MIRVs[42] and eventual countermeasures.[43] In addition, China will probably possess about two dozen DF-31 and CSS-3 ICBMs capable of reaching parts of the US. China is also developing a new and significantly more capable and survivable class of Type 094 SSBNs, which, together with the JL-2 SLBM system, will be able to reach the continental US from Chinese territorial waters. China's submarine-based strategic deterrence after 2010 is expected to consist of three to six SSBNs, each carrying 16 SLBMs.[44]

Even though this quantitative and qualitative development of strategic nuclear weapons systems will improve the Chinese second-strike capability, the surviving Chinese ballistic missiles will continue to be confronted by MD which must subsequently be penetrated. One analysis has estimated that 12–25 interceptors will be able to reduce a future Chinese ballistic missile force of between 102–190 warheads by about 20 per cent. Based on these assumptions, China can at most permit the deployment of about 20 interceptors in Alaska.[45] If the US deploys up to 100 interceptors in Alaska and probably increases this number further, MD will be able to absorb all the warheads in the planned Chinese modernization programme. Even though the interceptors' kill probability cannot be expected to be 100 per cent, it is doubtful whether Beijing for political reasons can permit itself to assume anything else than this worst-case scenario of a perfect missile defence. Under the preconditions of this analysis, which also does not include the Bush administration's plans to deploy sea-, air- and space-based intercept systems, China would be compelled to significantly increase the number of warheads in its ongoing modernization.

In connection with its ongoing modernization programme, China is thought to be pursuing several possibilities to counter MD. The Chinese efforts can be expected to include several different options, including a further increase in the number of ballistic missiles, deployment of MRVs or MIRVs, research and development of advanced penetrating countermeasures, such as decoys, stealth technology, electronic jamming, fast-burning boosters and manoeuvrable warheads, and direct countering of the most vulnerable parts of MD with anti-satellite weapons.[46]

China is assumed to have adequate amounts of fissionable materials to be able to increase its nuclear weapons arsenal to between 600 and 900 warheads, and as long as it chooses to increase the total number of warheads further, it will not be subject to any limitations in this respect over the long term. Other analyses assert that China will probably be able to produce up to 2,700 warheads.[47] China is further estimated to have had the capacity to develop MRV/MIRV systems for several years, and the US intelligence community assesses that China will be able to develop MRV to CSS-4 within a very few years.[48] China's attempts to provide MRV/MIRV capacity to the new generations of SLBMs and mobile ICBMs, however, is estimated to be extremely costly, just as it will entail very large technological difficulties.[49] This is partly because the new Chinese missile systems do not have the same payload capacity as does the current CSS-4 system, which makes them less suited for deploying several warheads, and even though the DF-41 might be provided with up to three MIRVs, it would impose very great technological demands in connection with the miniaturization of the warheads. In the same fashion, the development of manoeuvrable warheads, one of the countermeasure technologies emphasized as a Chinese option, also poses extreme demands on the design of the warheads. This can necessitate a revival of the Chinese nuclear tests, which might harm China's reputation in the international system and create problems for China in connection with the ratification of the CTBT treaty.[50]

In the 'Cox Report' of 1999, China was accused of conducting comprehensive and systematic espionage against American nuclear weapons programmes. Some American analysts assert that, if true, this would mean that the Chinese could more rapidly overcome several of the technological challenges connected with the design of advanced warheads.[51] The Cox Report concluded that China has acquired markedly improved possibilities to speed up the modernization of its nuclear weapons programmes, including especially the development of mobile ICBMs, SLBMs and MRV/MIRV capability.[52] Furthermore, the Cox Report hinted that China might also have acquired American computer programs to simulate nuclear tests.

China's opportunities to balance the combination of an American counter-force and MD could in principle also partly be realized by a 'hair-trigger alert'[53] doctrine. Confronted with an emerging crisis, China would then undertake the massive launching of all its ICBMs before these are destroyed in their silos by American missiles. A hair-trigger alert doctrine is problematic for three reasons, however. First, the Chinese ICBMs are maintained at a very low level of launch readiness, in that the missiles are not filled with liquid fuel, just as the warheads are removed from the missiles.[54] Second, China does not possess a BMEW[55] system. A previous Chinese attempt to purchase BMEW systems from Israel was probably blocked by the US following political pressure on the Israeli government.[56] Third, a launch-on-warning strategy can be viewed as a realistic option only in so far as a Chinese full-scale attack could be expected to actually penetrate MD. As long as this is not probable, China will at a minimum be compelled to improve its penetration ability by deploying more warheads and/or countermeasures. With its existing ICBMs, China will not be able to implement an adequately effective or credible launch-on-warning strategy.[57]

Regardless of which measures Beijing chooses to counter MD, it is of critical importance that at least some of these become clearly visible to the US. It is therefore extremely doubtful whether China will base its nuclear deterrence solely on uncertain estimates of both the effectiveness of MD and its own countermeasures. Even though Beijing might end up acknowledging that MD in practice will never be perfect, and that China will always be able to penetrate MD with the application of various countermeasures, Beijing will endeavour to create a powerful second-strike capability, consisting of a larger number of warheads in survivable ICBMs than America's MD could possibly be capable of engaging.

Despite its rapidly growing economic potential, China's possibilities of countering MD via an improvement of the strategic nuclear weapons' survivability could be limited by the need for domestic economic reforms. Hence, China needs a relatively cheap countermeasure against MD. At the same time, however, it needs a countermeasure that does not threaten other states, thereby creating the kind of regional instability that would ultimately threaten China's internal reforms.

China's response: regional implications

China's efforts to counterbalance MD will undoubtedly have regional security implications in East and South Asia; four regional issues, to be discussed in turn, are (1) the Chinese–Russian alliance, (2) the proliferation of weapons technology, (3) the 'North Korean card' and (4) the China–India–Pakistan balance.

The Chinese–Russian alliance

Beijing's attempts to counter MD via external balance have centred upon several joint diplomatic efforts with Russia in an attempt to maintain the now abandoned ABM treaty[58] in one form or another. China and the US have not signed any bilateral strategic arms control agreements. However, although China was not part of the ABM treaty, this has not kept Beijing from officially warning the US against withdrawing from the treaty, referring to its importance for the global nuclear balance of power.[59] China's interest in the ABM treaty stemmed from the fact that the treaty imposed on the US a wide range of limitations regarding MD which would help reduce the efforts needed by China to balance American nuclear strength.

Another very important diplomatic initiative has been China's entry into a bilateral treaty with Russia on 16 July 2001, a treaty that would lead to closer strategic cooperation between the two countries.[60] China and Russia have already concluded formalized cooperation arrangements via the Shanghai Cooperative Organization,[61] and the new treaty will contribute to strengthening what has until now been a rather loose but rapidly developing regional security structure. Both China and Russia have emphasized that the Sino-Russian agreement is based on a 'new type of interstate relations' that will create the basis for neither a military alliance nor a confrontation policy directed towards a third country.[62] However, the treaty's Article 9, which obligates the parties to immediately conduct consultations with each other in order to eliminate a perceived emerging threat, is interesting to note: it is not on a par with the collective defence obligation of the Washington Treaty's Article 5 but it does concern such an obligation.

The treaty must to a great extent be seen as grounded in the collective Sino-Russian resistance to an American missile defence. Several of the treaty articles refer directly to the maintenance of the global strategic balance and stability and to the associated agreements, primarily the ABM treaty.[63] The bilateral Sino-Russian relationship remains loose and weakly institutionalized, although some analysts believe that it has the potential to evolve into a stronger alliance over the long term.[64] The fact that the treaty is intended to be valid for an initial 20-year period with the possibility of a subsequent extension every fifth year also indicates that the bilateral cooperation can obtain more wide-ranging consequences. The key question may well turn out to be the mutual obligation towards strategic stability: China will likely argue that Russia must work to maintain China's level of nuclear deterrence towards the US at its current state,[65] but it is not clear that Russia is willing to go this far in its strategic relations with the US.

As long as the Sino-Russian cooperation comes to include research and development of advanced countermeasures and warheads, the treaty could also provide China with indirect access to Russia's significantly more developed strategic infrastructure.[66] By achieving both political and technological support for the modernization of its nuclear weapons forces, China would probably be able to gain most from the bilateral cooperation. An added benefit are the economic advantages to China which can also become a driving argument for a further expansion of Sino-Russian military cooperation; up to now these economic benefits primarily involve Russian arms exports to China.

Nuclear and missile technology proliferation

A second Chinese possibility to balance American deployment plans using 'external balance' takes its point of departure in a limited Chinese willingness to cooperate, or even conscious obstruction, within several areas of significant American security policy interest. In the threat perception of the US, China already occupies a key role as a major exporter of nuclear and missile technology to rogue states such as Iran, North Korea and Libya.[67]

China's export of missile technology to Pakistan has encountered vocal criticism from the US, which accuses China of having exported completely assembled M-11 missiles, missile components, missile-related technology and production assistance to Pakistan. The exports worry the US because they affect the fragile regional balance between the two rival nuclear powers – India and Pakistan – and because Chinese missiles and missile technology might be transferred to third counties, which can further destabilize international non-proliferation efforts. The US will be especially disturbed by the spread of nuclear weapons to other regions where it has vital strategic interests, i.e. primarily the Middle East and the Persian Gulf, where Iran, Syria and Libya occupy special roles.

In this context, the Chinese assessment that MD will result only in a greater proliferation of these technologies can be viewed as an unveiled

Chinese threat: as long as the US continues its build-up of MD and spread of TMD systems in East Asia, Beijing will feel less encouraged to ratify several international arms control and non-proliferation agreements.[68] In this connection, it is especially the MTCR[69] and NPT[70] which are of special significance for American non-proliferation efforts, and the US has exerted – and continues to exert – great efforts to keep China from exporting missiles, components and missile technology to Pakistan. Washington has used diplomatic pressure, offers of economic investment and economic sanctions in its attempt to bring China in under the MTCR regime.

Beijing has argued that China's membership of the MTCR must necessarily be seen in a broader context, which includes American arms sales to Taiwan and the development of TMD in East Asia. Over the long term, a continued or even intensified spread of technologies covered by the MTCR and NPT may even come to entail the export of advanced counter-measure technologies, including export to rogue states; this would complicate the American missile defence plans and perhaps even cause these to appear as ineffective and useless. In this context, however, Beijing can see itself confronted with a choice between initiatives, which on the one hand can be economically profitable and effective, but on the other hand can lead the US into the further intensification and development of its missile defence programme. As part of the modernization of its own strategic nuclear weapons, Beijing can also see itself pressured to postpone the ratification of the CTBT treaty and to withdraw from negotiations on the FMCT treaty,[71] all of which would have a negative influence on China's reputation in the international system.

The North Korean card

The US has repeatedly labelled North Korea a rogue state, and the prospect of North Korea developing both nuclear weapons and long-range ballistic missiles has been one of the key arguments for an early deployment of MD. In 2001, the US intelligence community concluded that by 2015 the US will probably be confronted with an ICBM threat from North Korea.[72] The fact that North Korea's missile and nuclear arms programmes already constitute a genuine threat to American allies in the East Asian region was demonstrated in August 1998, when North Korea test launched a 'Taepo Dong I' missile directly over the Japanese main island of Honshu. North Korea, to the profound dissatisfaction of the US, is also a major exporter of missile technology and has exported 'No Dong' missiles to Pakistan.

Since the Korean War, North Korea and China have been strategic partners, countering the alignment of the US, South Korea and Japan. Beijing's support of Pyongyang must be seen as grounded especially in the Chinese intention to maintain North Korea as a buffer state between

China and the American military forces in East Asia. From a political and military strategic perspective, however, China's geographic proximity to a rogue state is becoming a problem. In consequence, Beijing has several incentives to slow down or terminate the North Korean missile and nuclear weapons programmes.

First, a nuclear disarmament of North Korea would remove one of the most significant American arguments for MD. North Korea is the rogue state with the most advanced missile and nuclear weapons programmes, having also exhibited the most aggressive political intentions towards the US, for instance in the wake of the 2003 war in Iraq. A Chinese effort to retard or terminate North Korea's development of nuclear weapons would strengthen Beijing in its talks with Washington, and it would allow Beijing to argue that MD ought not to have an East Asian focused architecture, an argument Moscow likely would support.

Second, Chinese experts have repeatedly expressed the view that China desires a nuclear-free Korean peninsula. China fears that North Korean nuclear weapons would ultimately end up in the hands of a reunified Korea, which under South Korean leadership would direct these weapons against China. In a worst-case scenario, Beijing fears that the US would intervene militarily in North Korea with the intention of nuclear disarmament, and that such an intervention could result in a permanent American presence on the Korean peninsula. The US intervention in Iraq, which was ostensibly conducted with just this intention of disarming Iraq's weapons of mass destruction, makes such a threat real to Beijing.

Third, the presence of North Korean nuclear weapons could affect the dynamics of nuclear arms in the entire East Asian area, and stimulate other actors in the region to also acquire nuclear weapons. In particular, Beijing would expect that Japan, South Korea and Taiwan would begin to reassess their nuclear capabilities. Japanese nuclear armament would give sustenance to an already marked Chinese distrust of Japan's intentions and feed China's fear of a revival of Japanese militarism and imperialism. Finally, North Korean nuclear weapons also encourage China's neighbours to seek further American security guarantees, i.e. nuclear deterrence.

Fourth, the North Korean missile and nuclear weapons programmes are an important part of the American argument for cooperating with South Korea, Japan and Taiwan in the development and deployment of TMD systems in East Asia. In this connection, China will be anxious about the prospect of an integration of East Asian TMD systems into MD. An integrated TMD/MD system will not only be able to protect the US and its allies in East Asia against a nuclear threat from North Korea, it will also correspondingly reduce China's nuclear deterrence towards these states.

One can thus argue that China, despite the oppositely directed political and strategic motives, shares a common interest with the US and its allies in East Asia in preventing North Korea from possessing advanced missiles and nuclear weapons. China therefore occupies a very proactive role in the diplomatic efforts to convince North Korea to renounce its nuclear weapons and missile programmes. At present, China's interest is clearly indicated in the ongoing six-nation negotiations involving North Korea, the US, China, Russia, Japan and South Korea, negotiations now being conducted under Chinese auspices following the North Korean withdrawal from the NPT in January 2003. Still, three rounds of negotiations have not been able to convince North Korea to drop its weapons programmes, and China, despite its proactive role, has been criticized for not exerting sufficient pressure on its dependent neighbour.[73]

Both a temporary halt or an absolute termination of North Korea's nuclear weapon and missile programmes would support essential Chinese security interests, and Beijing has therefore a political and military incentive to replace its former strategic partnership with North Korea with a decisive and non-threatening distancing from Pyongyang. China strives to appear as a responsible international great power, which includes having good relations especially with the US, South Korea and Japan; these countries are also significant economic investors in China. In addition, Beijing might buy time to carry out its ICBM programmes at a pace suited to its technological and economic development.

North Korea has thus become a 'card' that China can 'play' politically against the US and its allies in East Asia, and Beijing wants to demonstrate its 'positive' influence on Pyongyang. North Korea is not an easy card to play, however: Pyongyang's resilience and veiled reasoning may raise the costs for China and thus reduce the cost-effectiveness of this card. This is also the likely reason why Beijing has tended to dismiss North Korea in its dealings with the US, arguing merely that North Korea does not constitute a genuine threat to the US, much less a credible justification for MD.

The China–India–Pakistan balance

Regardless of which internal countermeasures China chooses to offset the MD, Beijing must seriously consider the consequences of these measures for the balance of power and the dynamics of nuclear weapons in South Asia. As long as China endeavours to develop solely passive and/or active MD countermeasures, such countermeasures will be viewed only as a threat by states with missile defence, in this case, the US and its allies with TMD. However, if China chooses to deploy several missiles and warheads, this counter-move could also be viewed as a threat to the security of other states. Several analysts point out that the US deployment of MD will release a cascade effect in South Asia, in that India, as a response to China's expected military counter-move, might increase its nuclear armament, which would in turn lead to a corresponding rearmament in Pakistan.[74]

The dynamics of nuclear weapons in the South Asian region are regulated primarily by the bilateral balances of power between China and India and India and Pakistan. The nuclear weapons strategies of China, India and Pakistan are all principally based on counter-value doctrines, i.e. minimalist nuclear deterrence,[75] and in this context, the current qualitative and quantitative differences in the three countries' nuclear arsenals mean nothing or only very little. Several factors, however, can affect the dynamics of nuclear weapons and contribute to nuclear instability. First, none of the three countries' nuclear weapons and missile programmes is subject to any form of treaty-related limitations. Second, the nuclear weapons and missile programmes of all three states are characterized by a lack of transparency. And third, none of the three countries has any effective BMEW systems. In addition, there is neither quantitative nor qualitative parity between their strategic arsenals, and a continuing export of missile and nuclear weapons technology and missile defence systems to the region still occurs.

Since the 1960s, China and Pakistan have had a solid strategic alliance, which was originally grounded in a mutual need to balance the security threat from the Soviet Union and India. Today, the alliance continues to serve their security interests: Beijing uses its close ties with Islamabad to exert influence in South Asia and to balance against India; Pakistan needs a strong ally to balance India. China's export of missiles, components and missile technology to Pakistan occupies a key role in the two countries' alliance cooperation, and is also used by Beijing as a political 'lever' towards the US in the argument against the proliferation of TMD.

In sum, there are several strategic, economic and foreign policy rationales behind China's missile export to Pakistan. Nevertheless, China might be moved to cease its export activity. Such a decision could improve China's international reputation, better its relations with India and strengthen the credibility of China's critique of American proliferation of TMD in East Asia. After the expected modernization of China's strategic nuclear weapons, Beijing might also conclude that China does not need Islamabad to balance India to the same extent as earlier and that the Chinese missile export to Pakistan only exacerbates the cascade effect, thereby confronting China with an even stronger India.

For decades, China and India have been strategic rivals: the territorial conflict that led to the border war between China and India in 1962 remains unresolved, and the conflicts over Tibet (Xizang), Kashmir and Sikkim further complicate relations. China's strategic nuclear weapons therefore constitute a natural benchmark for India, and Beijing's response to MD will provoke concern in New Delhi and might thus stimulate renewed competition in South Asia. Some analysts assert that India has long accepted a quantitative and qualitative gap in relation to China's strategic nuclear forces, and that a further widening of this gap will not mean anything in itself as long as China remains vulnerable to India's nuclear retaliation.[76] However, it is important whether China continues to maintain its nuclear strategy, i.e. defensive minimalist deterrence, or whether Beijing decides to pursue a more clear-cut nuclear war-fighting capability. A modernization of China's strategic forces to deal with MD will not spawn an arms race with India as long as New Delhi remains convinced about the credibility of its nuclear deterrent vis-à-vis not only China but also China's protégé, Pakistan. To maintain its conviction, New Delhi is modernizing its nuclear weapons and missile capabilities. In August 2004, India test launched for the third time the land-mobile Agni-II missile, with a range of 2,500 km. Also in 2004, India is expected to test the Agni-III missile, which, with its 3,000–3,500 km range, can target major Chinese cities.[77]

Convictions are not based only on national efforts, however, but also on rivals' intentions. India must in this respect be particularly concerned with Pakistan and the territorial conflict over the province of Kashmir, which has resulted in four wars: in 1947, 1965, 1971 and most recently in Kargil in 1999. While these conventional wars could not be deterred, nuclear deterrence apparently worked in 1999. However, reports indicate that the short and bloody Kargil conflict came perilously close to evolving into a nuclear war, causing some analysts to question the deterring effect of nuclear weapons in South Asia.[78] Instability may therefore result from nuclear armament in the region. However, because nuclear deterrence in South Asia is based on minimalist counter-value strategy, a more likely threat to stability comes from the deployment of TMD. Faced with MD, India or Pakistan can be brought into a security dilemma, where especially non-transparency, the absence of conflict-preventive measures and lack of BMEW systems can encourage one of the parties to strike first with nuclear weapons in the case of a crisis.

Both India's and Pakistan's relatively small number of short- and medium-range missiles will thereby be vulnerable to TMD systems, which in this role will fulfil a distinctly national defence mission. Hence, India has already considered deploying missile defence systems and in this connection has expressed interest in American (PAC-3), Russian (upgraded S-300) and Israeli (Arrow) TMD systems. If India deploys TMD, Pakistan will undoubtedly also intensify its efforts to procure a corresponding system or will initiate other counter-moves. The potential for disruption between Pakistan and India is, however, reduced by the close involvement of the US with both countries: the US has made Pakistan a 'major non-NATO ally' in direct response to the war on international terrorism, and relations with India have generally and gradually improved since India conducted nuclear tests in 1998. Strategic stability between Pakistan and India is in US interests not only to avoid a war between the two countries but also because strategic cooperation between these countries and the US enables a type of 'soft containment' of China. Thus, stability might prevail in South Asia after all, but it will be a double-edged sword for China and might not contribute to the easing of Sino-American relations overall.

Conclusion

China finds itself compelled to alter both the developmental direction and the speed of its ongoing missile and nuclear weapons programme, perhaps even significantly beyond that which is already planned. China will be encouraged to assume a worst-case scenario for MD architecture and effectiveness, and will therefore be compelled to deploy at least as many warheads in survivable ICBMs as the number of interceptors in MD. In addition, China can be expected to develop and deploy a number of advanced countermeasures, which can improve the survivability of its warheads. The need for change in the Chinese modernization programmes can have economic implications that can threaten China's domestic reforms. China can therefore see itself compelled to choose the cheapest possible counter-move to MD. As a consequence, China may probably be content with being able to maintain a minimalist nuclear deterrence towards the US.

China's joint efforts with Russia to counter MD through external balancing have not impeded the American plans, and the alliance with Russia has yielded no visible political and military benefits to China. China's current use of the 'North Korean card' may be expected to have a stabilizing effect in the East Asian region. North Korea, however, is not the US's only justification for MD, and at most China can hope that the 'North Korean card' can have a delaying or reducing effect on MD.

Seen in isolation, China's response to MD will not necessarily have regional implications in South Asia. All three actors have already initiated the modernization of their strategic nuclear weapons, and the need for economic reforms can encourage the parties to maintain minimalist nuclear deterrence, which will inhibit a nuclear arms race. If the nuclear strategies are maintained, the balance of power in the region will be maintained, but at a higher level, and as long as the three actors do not pursue counter-force capability, i.e. missile defence, precise missiles, and so on, stability will not be threatened.

Continued or intensified spread of Chinese missiles and missile technology to Pakistan can eventually become a security problem for China if this leads to nuclear instability in South Asia. By dissolving its military alliance with Pakistan and at the same time halting proliferation, China might expect to achieve political and strategic benefits: its southwestern flank will be stabilized, and concessions might be extracted from the US in return for its increased ability to develop the partnership with Pakistan. This turn of events may take place, and the negative regional consequences of MD will thus be greatly reduced. Still, the fact of the matter is that the core question in Asia concerns the bilateral Sino-American relationship, and neither is likely to abandon its regional ambitions. In the short run we should therefore expect China to focus on its national capacity to counter MD; in the long run we might expect regional ripple effects and greater instability.

Notes

1 The common designation for states that constitute a special security threat to the US. The rogue states originally comprised North Korea, Iran and Iraq (the 'Axis of Evil'), but Syria and Libya have also occupied roles in this context.
2 George W. Bush, speech by the President at the National Defense University, 1 May 2001, Washington, DC: The White House; and George W. Bush, speech by the President in the White House, 13 December 2001, Washington, DC: The White House.
3 Intercontinental ballistic missiles.
4 Radar that can identify, track and discriminate between ballistic missiles, re-entry vehicles, countermeasures and other objects in the early phases of an ICBM's ballistic trajectory.
5 The US plans to field the so-called Initial Defensive Capability (IDC) in 2004–5. In October 2004 a total of six interceptors were deployed at Fort Greely on Shemya Island. In addition, two interceptors were planned for emplacement later in 2004 at Vandenberg Air Force Base in California.
6 The point of departure is Stephen Walt's neorealist 'balance of threat' theory, according to which states will react primarily to imbalances in their threat perception. Walt identifies four sources of threats: (i) aggregate power, (ii) geographic proximity, (iii) offensive capability and (iv) aggressive intentions. He argues that states will seek to balance an emergent threat by either intensifying their own internal efforts (internal balancing) and/or by entering into alliances (external balancing). Stephen M. Walt, *The Origin of Alliances*, Ithaca, NY: Cornell University Press, 1987, pp. 21–28 and 263–264. The analysis is further based on Robert Jervis's theory of 'Four Worlds under the Security Dilemma', which explores the consequences of the ambiguity between offensive and defensive weapons systems. Jervis argues that the security dilemma operates at full force in the worst-case scenario, i.e. when no differentiation can be made between offensive and defensive weapons, and when the offensive weapons have the advantage over the defensive weapons. Robert Jervis, 'Cooperation under the Security Dilemma', *World Politics*, 30, 1978, pp. 211–214.
7 Sha Zukang (a), speech at NMD briefing, 14 March 2001, Beijing: Foreign Ministry of the People's Republic of China.
8 Mostly members of the Falun Gong movement and militant Uighurs in China's western Xinjiang province, who are fighting for an independent East Turkestan.
9 Douglas T. Stuart and William T. Tow, *A US Strategy for the Asia-Pacific*, Oxford: Oxford University Press for the International Institute for Strategic Studies, 1995, pp. 6–27. Most important in this connection are US relations with Taiwan, which are formalized in the Joint Defense Treaty from 1954, the Taiwan Relations Act from 1979 and the Taiwan Security Enhancement Act from 1999. Japanese–US cooperation is formalized via, among other things, the US–Japanese Defense Cooperation Guidelines, revised in 1997, and the Treaty of Kanagawa (Japan–US Treaty of Peace and Amity), signed on 31 March 2004.

10 Dingli Shen, *What Missile Defense Says to China*, Shanghai: Fundan University, Center for American Studies, 2000; Abram N. Shulsky, *Deterrence Theory and Chinese Behavior*, Santa Barbara: Research and Development Organization, 2000, p. 17.
11 Shulsky, op. cit., p. 19.
12 Charles Ferguson, *Sparking a Build-up: US Missile Defense and China's Nuclear Arsenal*, Washington, DC: Arms Control Association, 2001; Dean A. Wilkening, *Ballistic Missile Defense and Strategic Stability*, Oxford: Oxford University Press for the International Institute for Strategic Studies, 2000, p. 21; Shulsky, op. cit., p. 21.
13 Wilkening, op. cit., p. 21.
14 In 1999, two of President Bush's current security advisers, Richard Perle and Deputy Secretary of Defense Paul Wolfowitz, together with 23 other leading conservative politicians, signed an anti-Chinese declaration asserting that China continues to threaten to apply military force against Taiwan's democratically elected government. The group called on the US to make an all-out effort to deter every form of Chinese intimidation of Taiwan, and for the US to declare its unequivocal support for the defence of Taiwan in case of either an attack or a blockade against Taiwan or the islands of Matsu and Kinmen. Caspar Weinberger, R. James Woolsey and William J. Bennett contributed with their signatures to give the declaration a certain weight. The Heritage Foundation, *Leading Conservatives, Foreign-Policy Experts Call for Defense of Taiwan*, Washington, DC: The Heritage Foundation, 1999.
15 See Ministry of Foreign Affairs of the People's Republic of China, *The Taiwan Question in China–US Relations*, Beijing: Foreign Ministry of the People's Republic of China, 2000.
16 Ibid.
17 Theatre missile defence.
18 *The China Daily*, 19 July 2001.
19 Zukang (a), op. cit.
20 Jim Garamone, *It's Not 'National' or 'Theatre', It's Just Missile Defense*, Washington, DC: American Forces Press Service, 9 March 2001.
21 Sha Zukang (b), briefing on the missile defense question, 23 March 2001, Beijing: Foreign Ministry of the People's Republic of China.
22 Ferguson, op. cit.; Charles S. Glaser and Steve Fetter, 'National Missile Defense and the Future of US Nuclear Weapons Policy', *International Security*, 26, 2001, p. 81.
23 Condoleezza Rice, 'Promoting the National Interest', *Foreign Affairs*, 79, January–February 2000, p. 56.
24 Peter Paret (ed.), *Makers of Modern Strategy: From Machiavelli to the Nuclear Age*, New York: Oxford University Press, 1986.
25 ICBMs, SSBNs and strategic bombers.
26 Nuclear-powered ballistic missile submarine.
27 Submarine-launched ballistic missiles.
28 Anti-submarine warfare.
29 *Counter-force* designates the capacity to fight the enemy's nuclear strategic forces, while *counter-value* is the ability to destroy the enemy's highest priority targets, i.e. population and industrial centres. See Lynn Eden and Steven E. Miller (eds), *Nuclear Arguments: Understanding the Strategic Nuclear Arms and Arms Control Debates*, Ithaca, NY: Cornell University Press, 1989, p. 14.
30 National Intelligence Council, *National Intelligence Estimate on Foreign Missile Developments and the Ballistic Missile Threat through 2015 (Unclassified*

Summary), Washington, DC: Central Intelligence Agency, 2001; Ferguson, op. cit.

31 Jiang Zemin, speech at the Conference on Disarmament in Geneva, 26 March 1999, Beijing: Foreign Ministry of the People's Republic of China; Zukang, (b), op. cit.

32 Ferguson, op. cit.; Zukang (b), op. cit.; Li Bin, *The Effects of NMD on Chinese Strategy*, Beijing: Tsinghua University, Institute of International Studies, 2001.

33 Mark A. Stokes, *China's Strategic Modernization: Implications for the United States*, Carlisle, Pa.: US Army War College, Strategic Studies Institute, 1999, p. 10; Ferguson, op. cit.

34 Bin, op. cit.

35 National Intelligence Council, op. cit.

36 Glaser and Fetter, op. cit., p. 82.

37 Ibid.

38 National Intelligence Council, op. cit.

39 Bin, op. cit.

40 National Intelligence Council, op. cit.

41 Multiple re-entry vehicles, i.e. missiles with several nuclear warheads.

42 Multiple independently targetable re-entry vehicles, i.e. missiles with several nuclear warheads which can be guided independently towards separate targets.

43 National Intelligence Council, op. cit.; Wilkening, op. cit., pp. 83–84. The latter source maintains that the estimate of China's future strategic nuclear forces is connected with a relatively large uncertainty as a result of the lack of information about modernization plans.

44 Ibid.

45 The conclusion of the analyses must be seen in light of the fact that the limit of 20 per cent is set as the probable number of Chinese warheads that would survive following a pre-emptive American nuclear strike. The hit probability of the interceptors is estimated at 90 per cent and a certain proportion of the Chinese ballistic missiles are assumed to have deployed either two or three MIRVs. See Wilkening, op. cit., p. 42.

46 Wilkening, op. cit., p. 19; Glaser and Fetter, op. cit., p. 63.

47 Wilkening, op. cit., p. 83.

48 National Intelligence Council, op. cit.

49 Bin, op. cit.; National Intelligence Council, op. cit.

50 Comprehensive Test Ban Treaty.

51 US Congress, *Report of the Select Committee on US National Security and Military/Commercial Concerns with the People's Republic of China*, Washington, DC: US Congress, 1999.

52 National Intelligence Council, op. cit.

53 'Hair-trigger alert' is another designation for 'launch-on-warning', a launching doctrine by which a retaliatory nuclear strike is launched at the warning of an enemy attack. 'Hair-trigger alert' poses great demands on the credibility of a BMEW system, and the doctrine is associated with a certain degree of risk of mistaken launching.

54 Bin, op. cit.; Glaser and Fetter, op. cit., p. 81.

55 Ballistic missile early warning, i.e. a radar system for detection of attacking ballistic missiles.

56 *The China Daily*, 19 July 2001.

57 Bin, op. cit.

58 Anti-Ballistic Missile Treaty.

59 Zukang (a), op. cit.
60 Ministry of Foreign Affairs of the People's Republic of China, *The Treaty of Good-Neighborliness and Friendly Cooperation Between the People's Republic of China and the Russian Federation*, Beijing, 2001.
61 Also called 'The Shanghai Six', which besides Russia and China also includes Kazakhstan, Kyrgyzstan, Tajikistan and Uzbekistan. The organisation, founded in 1996 by the first five of these countries, has the goal of initiating and supporting confidence-building measures in the border regions.
62 Foreign Ministry of Russia, *Joint Declaration on Behalf of the Russian and Chinese Heads of State*, Moscow: Foreign Ministry of Russia, 18 July 2000.
63 See Articles 11 and 12 of the treaty.
64 Nikolai Sokov, *What Is at Stake for the United States in the Sino-Russian Friendship Treaty?* Monterey: Monterey Institute of International Studies, 2001, p. 3.
65 Ibid., p. 2.
66 Ibid.
67 George J. Tenet, *Worldwide Threat 2001: National Security in a Changing World*, Washington, DC: Central Intelligence Agency, 7 February 2001; National Intelligence Council, op. cit.
68 Zukang (b), op. cit.
69 Missile Technology Control Regime.
70 Nuclear Non-Proliferation Treaty.
71 Fissile Material Cut-off Treaty.
72 National Intelligence Council, op. cit.
73 China is the largest exporter of energy and food to North Korea.
74 See among others Michael Krepon, *Missile Defense and the Asian Cascade*, Washington, DC: The Henry L. Stimson Center, 2002, p. 95.
75 Andrew Feickert and K. Alan Kronstadt, *Report for Congress: Missile Proliferation and the Strategic Balance in South Asia*, Washington, DC: US Congressional Research Service, 2003, p. 9.
76 Rajesh M. Basrur, *Missile Defense and South Asia: An Indian Perspective*, Washington, DC: The Henry L. Stimson Center, 2002, p. 13.
77 'India Planning to Test Agni-III Missile', *Defense India*, 4 June 2004.
78 Feickert and Kronstadt, op. cit., pp. 8–9.

7

THE MIDDLE EAST AND MISSILE DEFENCE

Birthe Hansen

Introduction

The Middle East[1] has so far played a pivotal role in the debate on US missile defence (MD) for at least four reasons. First, the region was the first to gain experience with the use of a missile defence system when Iraq fired Scud missiles against Israel in January 1991, and the Patriot system was activated. Second, the region holds the first regionally developed missile defence system (although much less ambitious than the American system). Israel's Arrow II system[2] has been tested, and the tests have proved comparatively successful. Third, the Middle East is a conflict-prone region and priority to programmes of weapons of mass destruction (WMD) has been given since the 1960s. Later, indigenous development of ballistic missiles has taken place.[3] Fourth, the region has been an important part of the US reasoning for developing MD: the US has put an emphasis on Middle Eastern WMD and missile programmes in combination with the so-called rogue policy of some of these states when formulating its official positions.

Given these reasons, the Middle East has become an important and integrated part of the debate on the US MD. The main question is how the US plans and the Middle Eastern development will interact: will the bad scenario prevail, in which US plans spur further regional armament, conflict and world order defiance? Or will the good scenario prevail, in which the plans spur regional disarmament, temper the conflict potentials and provide an acceptance of the world order instead? These two positions have emerged in the debate on the US MD since the late 1990s when the Clinton administration began to plan deployment after pressure from Republican members of Congress.

This chapter[4] argues that we cannot exclude the long-term prevalence of either of the scenarios, but that the US policy in combination with the regional dynamics seems to favour the good scenario.

Of course, the conclusive answer to this question cannot be given yet, as we are interested in the long-term effects. Planning countermeasures, developing strategies, acquisition policies, and the development of weapons take time and require resources. Indeed, this is of relevance in the case of the Middle East, because the region is currently in a weak position in respect to most capabilities, and it therefore lacks the necessary resources to fully balance or oppose the US initiative.[5] Furthermore, it is difficult to sort out the effects in a region that is subject to change and turbulence, and which is therefore influenced by many other factors than the US MD plans.

However, we can analyse the background, the policies of Middle Eastern states, the US strategy, and the regional logics in order to provide a platform for discussing the interaction between the MD plan and the Middle Eastern development. The four realms are chosen because they affect the US missile plans as well as the Middle Eastern responses. Consequently, this analysis is divided into four sections:

1 The background: regional trends regarding conflict, WMD, missiles and terrorism.
2 The policies of Middle Eastern states: their priorities and current challenges.
3 The regional strategy of the US.
4 The logics: defence concerns, strategies of ambiguity, and political choices between armament or world order adjustment.

Following the analysis, the conclusions are drawn that the US MD might contribute to affecting Middle Eastern security in a positive way, but only if the MD is accompanied by a series of other measures in order to limit the proliferation of WMD and ballistic missiles, and to reduce the regional conflict potential.

The analysis is based on the assumption that the US as a unipole is seeking to be on the edge in general. The missile defence project is seen as 'the edge' within the military realm, as it represents a possible first-strike capability and reflects military superiority.[6] Precisely how it is going to develop is less important for the analysis than the project in itself.

Both the scenarios are analysed in each of the four sections according to two sets of assumptions. The good scenario is assumed to depend on the rising cost of pursuing WMD and missile strategies, and on a strong US position, which encourages adjustment to the current world order. The bad scenario is assumed to depend on state panic and cornered positions, and on the involvement of failed states and 'revolutionary' regimes, which produce a risk of conflict and cornered positions.

Background: regional trends and experiences

The Middle East has been ridden with conflicts, the states have been searching for WMD, WMD have actually been used in the region, and now also terrorists are on the lookout for a capacity. These four regional characteristics have been part of the US reasoning for planning the MD, but they are also important regional features, which affect the behaviour of the regional states. To both the US and the Middle Eastern states, the four characteristics are tightly connected to the security dilemma. They fuel distrust, lack of confidence and a reliance on military capabilities.

The conflict-ridden background is clearly revealed by listing the series of regional conflicts (war, civil wars, substantial upheavals), that have taken place in the first 15 years after the end of the Cold War in 1989:

- The Iraqi invasion of Kuwait (1990)
- The Kurdish and Shiite upheavals in Iraq (1991)
- The civil war in Algeria in the 1990s
- The civil war in Afghanistan in the 1990s
- The Yemeni civil war (1994)
- The Palestinian upheaval (2000–)
- Clan fighting in Afghanistan (2002–)
- Upheavals and unrest in Iraq (2003–)

In addition to these conflicts, the region has been characterized by the presence of a series of unresolved conflicts and border issues. Furthermore, the region has increasingly become the target of terrorism itself as well as a supplier of international terrorism.

Although no WMD have so far been found in Iraq after the 2003 war, it is an uncontested point that WMD programmes have been part and parcel of the regional development since the 1960s. Around that time, a virtually spiralling development of chemical weapons took off. Egypt was probably the first state to acquire a chemical capacity, and it was also reported to be the first state to use CW – against Yemeni royalists in the 1960s.[7] The UN weapons inspectors basically destroyed Iraq's CW capacity during the 1990s, but Iraq had used its capacity at least 14 times during the war against Iran prior to the destruction.[8] Iran developed chemical weapons in the late 1980s in response to the Iraqi capacity but resorted to only limited use.[9] Israel is considered to have achieved a chemical capacity in the 1970s, and so did Libya (which is supposed to have used CW against Chad in 1987)[10] and Syria.

Biological weapons also became an option. When the UN weapons inspectors examined Iraqi capacities after Operation Desert Storm, they found a biological capacity, which was much larger than had been previously estimated.[11] In addition to Iraq, Egypt, Libya, Syria, Israel and Iran initiated research into BW programmes.

Israel is the only nuclear power in the Middle East. While Israel has never admitted to having gone nuclear, there is a general analytical agreement that Israel probably has at least 100 warheads, possibly more than 200. Iraq, Iran and Libya have had nuclear programmes, and Iran is continuing its efforts. After the 2003 war against Iraq, Iran and the IAEA had several disputes on the findings of enriched uranium, which Iran argued was for civilian purposes. The IAEA, the US and several other countries, however, rejected the argument.[12]

In general the Arab WMD programmes were halted in the 1990s owing to lack of resources, the disappearance of Soviet economic and technical support, general regional decline, and the emerging peace process. The war on Iraq has contributed to the current decrease in the programmes by terminating the Iraqi efforts and scaring Libya into officially giving up.[13]

Research and development of WMD programmes, however, is one thing. The actual *use* of the weapons is another. Since the end of the Cold War, no use of WMD in the Middle East has been reported. Consequently, it might be useful to look at the pre-1989 cases in order to extract a possible lesson on the circumstances of their use. Iraq felt free to use CW against Iran during the war in the 1980s. Likewise, Libya and Egypt reportedly have each used CW in a very limited way in minor, regionally isolated conflicts. In contrast to this, Iraq did not use CW (or BW) during Desert Storm in 1991 even though it was under strong pressure and about to lose the war. In the cases of use, the conflicts were local, and the superpowers did not engage. In 1991 Iraq was facing a huge and credible threat of retaliation – both in terms of capacity and willingness to do so. At the time, US Secretary of State James Baker delivered such a threat to Iraq during last-minute negotiations in Geneva.

Thus, the lesson from these very few cases is: states with a WMD (CW) capacity are inclined to use it in local conflicts if they 'need' to and feel free to do so, but they are less likely to use WMD if they face a substantial risk of retaliation.

In the post-Cold War era, concern over the use of WMD has been extended to terrorist groups. Particular concerns arose after the 1995 Aum Shinrikyo attack in Japan. Twelve people were killed, and thousands were injured, when the extremist religious cult released the chemical agent sarin in the Tokyo subway. The threat was further underlined by 9/11. These attacks, which followed a series of al-Qaeda terrorist attacks in the 1990s, demonstrated al-Qaeda's ability to carry out mass murder even by means of conventional weapons. Later, Osama bin Laden declared that it was a duty for al-Qaeda to use nuclear weapons, and in Afghanistan, indications about al-Qaeda's WMD ambitions were found in the aftermath of Operation Enduring Freedom.

It has often been emphasized that WMD are difficult to use, particularly BW, and that terrorist organizations would lack the ability to launch the warheads with missiles. However, in November 2002, presumed al-Qaeda terrorists almost hit an Israeli aeroplane with a shoulder-fired missile, when the plane took off from Mombasa Airport, Kenya.[14] The attack showed that a terrorist missile capacity was no longer a problem of the distant future.

It is evident that there are many other ways to carry out terrorist attacks, and that the current wave of suicide terrorism expands the range of alternatives. Nevertheless, the information of al-Qaeda plans shows that it would be irresponsible to rule out terrorist use of WMD, and the Mombasa incident likewise shows that terrorist use of missiles is an idea already born.

If we sum up the Middle Eastern background and trends regarding threats arising from regional conflicts, WMD, missiles and terrorism, and relate the findings to the US MD plans, we obtain the following picture:

- Conflicts with a risk of escalation are still part of the regional characteristics. Such conflicts increase the risk of parties feeling cornered and inclined to resort to the use of weapons available. This provides an incentive for the US to maintain its edge, and to develop instruments for dealing with regional conflicts, including measures to defend troops deployed abroad.
- The reliance on WMD programmes seems to have decreased during the regional decline of the 1990s, which drained the states of resources and ambition. The progress in missile technology seems to have suffered a backlash as well. Iraq was disarmed (and came under US control), and Libya officially gave up its programme in 2003. Thus the Middle East does not seem to be the 'first' reason for a US MD. On the other hand, regional defence needs are still urgent in the light of the conflict potentials, and the use of WMD has previously been comparatively high in the region. Only a nuclear superiority with a credible threat of overwhelming retaliation seems able to prevent the regional use of WMD. This provides the US with a strong incentive to preserve its edge in general.
- The rising threat from terrorism (and the current wave of terrorism has its origins in the region) points to the need for US awareness of many other instruments, but also of MD in the longer term.

The picture thus reflects that the US is provided with different incentives arising from the conflict-prone Middle Eastern background. However, preventing the use of WMD is presumably the strongest incentive, and this comprises the US need to be on the edge in order to maintain the necessary ability of a credible deterrence.

The policies and challenges of Middle Eastern states

Most Middle Eastern states are generally strongly opposed to the US MD project. They fear that an increased US military position will lead to US dominance in the region. Israel has a different perspective. Although not fully trusting its US ally, Israel is also facing an 'imperative', which is encouraging it to support the US efforts.[15]

Israel lacks regional allies, and it is still engaged in unresolved disputes (including the Palestinian problem). It therefore has an interest in a US ability to manage regional conflicts. Of course, if Israel and the US fell out, the interest would change, but Israel can hardly afford this in the foreseeable future. Less amicable relations with the US would require better regional relations, but better regional relations depend on US efforts.[16]

The Israeli–US cooperation in terms of regional missile defence has caused concern in Arab states hostile to either of the parties or both. In addition, internal Arab developments may affect the outcome of the equation. The Arab world is currently going through a wave of leadership succession. A series of Arab leaders came to power at the end of the 1960s and during the 1970s. Some have already been replaced while others are next in line – if for no other reason than age.

In a series of Arab countries, leadership succession is based on a mixture of principles, which holds the potential for power struggles and eventually the outbreak of armed conflict.[17] In Oman, for example, the childless Sultan Qaboos has given instructions that his family is to be consulted if he dies, and if they do not agree on his successor, the Omani Defence Council has to choose between two officially secret candidates, whose names Qaboos has written down and hidden in sealed envelopes.[18]

The challenge of Arab successions has to be seen in the light of the economic depression in the region, the decreasing oil reserves and the quest for democratization. These three factors hold the risk of adding to the potential for conflict. Of course, the US MD project will not in any way facilitate these developments, but it might be a piece of insurance in the case of new conflicts and prevent new leaderships from adopting offensive strategies of protection. In this way, the project may even assist the US pressure for democratization in the Middle East, which requires an adjustment to the world order rather than offensive strategies of protection from the demands of the world order.

The MD plans will not directly assist in reducing conflicts and skirmishes among the Middle Eastern countries, which are having a series of (cold) conflicts. However, if some states give up their previous offensive strategies of protection, this will probably ease regional negotiations as well. In particular, strategies for going nuclear affect not only the US but also neighbouring states and other relations.

The fall of the Saddam Hussein leadership in Iraq, and the decreasing number of probable possessors, according to Eitan Barak, provides a window of opportunity for the elimination of chemical weapons. Until 2003, five of the seven probable CW possessor states were located in the Middle East.[19] After the war on Iraq and the changed Libyan policy, only three states are considered still to be in possession of CW: Egypt, Iraq and Syria.[20] A successful elimination, however, requires an end to the Arab linkage approach between CW and NW disarmament.[21]

Following Barak's argument, the Middle Eastern problem related to the Israeli nuclear status will diminish in the case of CW disarmament, which would prevent Israel from fearing a WMD 'slippery slope' in the region. To avoid this, Barak suggested that 'Israel ratified the CWC, the CTBT, and pledged to join the future FMCT'. Thereby, trade-offs will be provided to the Arab states.[22] If these trade-offs connected to the CW disarmament are recognized, the Israeli nuclear status may prove a little more acceptable to the Arab states.[23] Israel, on the other hand, would avoid 'the slippery slope'.

The development of the MD system may contribute to support the trade-offs in the case of both the Arab states and Israel. It may serve to devalue the Israeli nuclear capacity vis-à-vis the Arab states, while it is still able to protect Israel by raising the costs of Arab nuclear ambitions in the future and thereby still compensate inferiority in numbers with superiority in deterrence. On the other hand, the bad scenario cannot be ruled out: a power struggle like the one we saw between Saddam Hussein's Iraq and the US could follow the MD-related fear of dominance. However, since the end of the Cold War, the Arab world has suffered a decline and is not particularly suited for engaging in such a struggle.

In sum, the range of possible Arab, Iranian and Israeli policy options is too broad to lead to a single conclusion on the impact of the MD in the Middle East. However, the post-Cold War Arab decline seems to favour adjustment strategies in the light of the war on Iraq and the rising costs of future offensive protection strategies in case of a credible MD system and US edge.

The US strategy

Initially it was assumed that the US efforts to develop and deploy a missile defence system relate not only to the specific properties of such a system, but also to preserving the US general military and unipolar edge.

This US ambition evidently has a double impact. In the first place, it provides the US with a strong managerial capacity, which might offset the regional reasons behind state strategies for achieving nuclear status.

In the second place, the ambition can hardly help to spur Arab resentment. The war on Iraq had by 2004 led to increased popular anti-American sentiments, and several Arab governments had expressed insecurity and criticism.

Once again, a dilemma arises: while the MD project will probably incite adjustment strategies (including democratization) instead of offensive strategies of protection, it will probably give rise to resistance as well. Increased US superiority, which fuels popular anger in the Arab world in a situation of regime change, may bring about leaderships hostile to the US. Dramatically different leaderships in previously authoritarian areas are typically inexperienced, and inexperienced leaderships risk creating misperceptions and overreactions by spreading universal goals (which are perceived to be offensive) and confusing signals.[24] Prominent cases were France and the Soviet Union after the revolutions; in the Middle East, Iran became an example after the 1979 revolution. The subsequent risk is that the states are attacked.

Consequently, processes of transition need to be accompanied by political attention and support. In addition, democracies tend to be most fragile in their 'childhood', in which checks and balances have not been established. Also, processes subsequent to democratic transition thus need to be carefully monitored.

The US devotes many resources to its MD project. In the light of the Middle Eastern challenges it certainly needs to devote resources to back political processes, too. The two most urgent challenges will be to secure new democracies and provide them with incentives to pursue adjustment strategies, and to contribute to manage the signals of inexperienced leaderships.

Finally, a long-term and less spectacular effort needs to be fully integrated in the US policy: the slow, troublesome and detailed effort to develop full-fledged non-proliferation regimes. In this context, one example is illustrative: the insufficient efforts to provide a verification regime regarding the prevention of proliferation of biological weapons.[25]

Arms control and non-proliferation regimes have often been criticized for delaying but not preventing proliferation. However, delay may be worth while, as rapid proliferation in principle creates the most dangerous situations. Kenneth Waltz argued, in his famous and controversial 1981 article, that more nuclear proliferation might be a better option than less, because the possession of nuclear weapons tends to make the politicians in charge more responsible and reduce the risk of war.[26] Yet Waltz also argued that rapid proliferation is extremely dangerous, as it encourages pre-emptive strikes in order to prevent the proliferation as well as leaving the nuclear newcomers without the necessary security measures. By 2004, the basic Waltzian argument on the positive effects of proliferation still holds for major powers. However, we have had negative experiences in the Middle

East with respect to the use of non-nuclear WMD, and terrorists have appeared on the international stage. Terrorists are actors different from states and governments, and except for nationalist terrorist groups, they do not have territories to defend. They are thus comparatively free to act in the era of proliferation, and a broad range of counter-proliferation measures are therefore to be preferred in addition to the cost-enhancing MD project.

Furthermore, the wave of succession in the Arab world holds promises in terms of democratization, but it also holds the risk of state collapses. Some failed states – or even dysfunctional states – have previously turned out to shelter terrorists and/or to provide convenient room for their operative efforts. The democratization process thus needs to be assisted in order to avoid non-state actors (with different incentives and different cost-benefit analyses) undermining an otherwise progressive development.

Finally, the Israeli–Palestinian question needs to be addressed. The Palestinian uprising, which broke out in the autumn of 2000, risks cornering the Palestinians and breaking down Israeli society. The conflict is also part of the general US diplomatic game in the Middle East, in which the war on Iraq in 2003 has left a US need for rebalancing its efforts. A series of additional measures are thus necessary to prevent regional developments from offsetting the positive effects of the US MD project.

The logics: regional incentives

The end of the Cold War implied that a series of Middle Eastern states and actors lost their superpower ally and important specific support from the Soviet Union. Instead, they had to choose between two alternative strategies: either to join the American world order or to protect themselves against it.

Having no major allies after 1989, these states had to rely on themselves when elaborating strategies of protection. This was a huge challenge, as they were all comparatively weak in capabilities. One way forward was to opt for a nuclear deterrent, which is the effective way of protection against invasions and strong pressure.

However, threshold states are prone to face the risk of pre-emptive strikes when getting close to achieving a nuclear capacity. This provides them with an incentive to pursue strategies of ambiguity as a variation of protection strategies. A strategy of ambiguity includes confusing signals, hints, half-hearted negotiations, and the rejection of openness regarding their nuclear efforts. If other states are in doubt as to whether or not the nuclear threshold has been crossed, they will be likely to act as if it were, and they will be less likely to engage in pre-emptive action. On the other hand, it becomes more difficult for the adversaries to mobilize international society or to legitimize warfare if no clear signal of an immediate ability to cross the nuclear threshold is evident.[27]

Two Middle Eastern states chose to rely on strategies of ambiguity: Iraq and Iran. However, the Iraq war of 2003 put an end to the Iraqi strategy. Furthermore, the search for WMD in post-war Iraq revealed that the Iraqi strategy had been based on bluff rather than on hidden progress.

The US MD plans could further either of two incentives regarding the future pursuit of strategies of ambiguity. On the one hand, it could raise the stakes and the quest for protection, while delaying the rhetorical dimensions of such strategies. States may panic or feel an even greater need to protect themselves and prevent further US intervention, and they may therefore pursue hidden programmes while keeping a low profile until they can actually cross the threshold. This indicates a development in line with the bad scenario.

On the other hand, the MD plans may offset the benefits of strategies of ambiguity. If a small nuclear deterrent becomes ineffective, or if the MD system becomes the symbol of an overwhelming capability (which holds the ability to deter even a nuclear retaliation strike), alternative strategies may prove to be favourable. That is, it may become too costly to seek to go nuclear in order to compensate for the lack of other capabilities, and then the cost-benefit analysis of the states in question will change. They will lose less by joining the world order and risking some US pressure than by pursuing offensive strategies of protection. This indicates a development in line with the good scenario.

Given the fact that the Middle Eastern states are weak in capabilities as the region in general has been subject to decline since the end of the Cold War, a US attempt to maintain and even increase its military edge seems to favour alternative strategies.

Iran is the only state that currently pursues a strategy of ambiguity. However, Iran is the state best suited for not changing direction and choosing an alternative strategy, as it is the strongest state (measured in terms of aggregate capabilities) in the Middle East, and it has probably already come a long way with its nuclear programme.

The US MD plans thus seem to encourage an end to WMD programmes in the Middle East – except in Iran. The Iranian case, on the other hand, strengthens the US incentive to carry on the MD project for post-crossing-the-threshold managerial reasons, as well as the Israeli case has done. This illuminates the difference between the MD plans and the US advantages of an operative and effective MD system: the plans may prevent proliferation, while an effective system will be useful to the US in managing efforts in respect to a (more or less) nuclearized region.

Conclusions: missile defence and the Middle East

Each of the four areas analysed above has shown that the US MD system is likely to affect the Middle East profoundly if implemented.

Likewise, the analysis has shown that the current and future Middle Eastern development provides the US with a strong incentive to carry on its MD efforts in order to work out an operative system and to preserve its military edge in general. The results specifically showed that the US plans for its MD system would interact with Middle Eastern politics in the following ways, measured according to the four areas in question:

- The record and probability of conflicts in the Middle East, the regional use of WMD, and empowerment of terrorist organizations with regional origins provide an incentive for the US to implement the MD.
- Regarding Middle Eastern politics, the analysis showed that the findings point in different directions. However, they tended to favour the conclusion that an extended US edge stemming from a credible MD system will encourage Arab states to pursue strategies of adjustment rather than protection – in the light of the rising costs of offensive protection.
- The US MD project will probably assist democratization processes in the Middle East and add to the level of inter-state security. This, however, requires a series of additional measures to support the complex processes.
- The MD plans will raise the cost of protection strategies and thereby favour adjustment strategies among Middle Eastern states – probably with the exception of Iran. Israel is considered a de facto nuclear power, and thus has a wider room for manoeuvre.

Among these partial conclusions, the most important is that the presence of a power in possession of an overwhelming ability to retaliate is a necessity in order to prevent the use of WMD in some regional conflicts. This naturally has several negative side effects, to be politically assessed. However, preventing the use of WMD can hardly be given too much priority.

In the light of these partial conclusions, we return to the two scenarios that have prevailed in the debate on the US MD. The good scenario in which the MD project (progressing successfully) makes it politically and economically too expensive to maintain existing Middle Eastern WMD and missile programmes seems likely in the light of the Middle Eastern material weakening and the democratic potentials of the wave of successions. The bad scenario in which the MD project makes states feel cornered, and failed or Islamist states result from the successions (and provide shelter for terrorist networks), cannot be excluded. Given the costs of pursuing such protective strategies and the international awareness of terrorism, this appears to be a less likely scenario than the good one – if the US maintains its relative strength. Another version is that the MD project will not progress successfully because the US is weakened (from the war on Iraq and its repercussions), or because the technological challenges have been underestimated by the US. In these cases, however, the major problem is not the MD project but the effects on the international system of a US that is weaker than expected.

The overall conclusion, thus, is that the US MD project – analysed as a US way of securing its military edge in general – will have a positive impact on the Middle East *if* it is complemented by support for democratization and additional non-proliferation measures. Of course, regional developments are influenced by a series of factors other than the MD, and the political game is complex. It will be overlaid by the fear of Middle Eastern states that the US is getting too much power, while their own strategic options are being limited. In particular, Syrian, Egyptian and Iranian leaders have articulated such fear, but in principle this is a concern for all the states.

Notes

1 Geographically, 'the Middle East' is understood as the region from Morocco in the west to Afghanistan in the east, Turkey in the north, and the Arabian Peninsula in the south.
2 A comprehensive analysis of the Israeli programme is provided by Cameron S. Brown, 'Israel and the WMD Threat: Lessons for Europe', *Middle East Review of International Affairs*, 8(3), September 2004.
3 Yiftah Shapir, 'Ballistic Missiles in the Middle East', *JCSS Bulletin*, No. 17–18, March 1997. Online. Available HTTP: ⟨http://www.tau.ac.il/jcss/bulletin/bull7p2.html⟩ (accessed 7 October 2004).
4 The author wishes to thank Carsten Jensen, Bertel Heurlin, Kristian Søby Kristensen and Sten Rynning for comments.
5 The Middle East was for many years the number one region regarding arms purchases. Recently, the decline has led to a number two position.
6 Birthe Hansen, *Nye våben i syd*, Copenhagen: DUPI, 2001.
7 E.J. Hoogendorn, 'A Chemical Weapons Atlas', *Bulletin of the Atomic Scientists*, 53(5), September–October 1997, p. 37.
8 Avigdor Haselkorn, *The Continuing Storm: Iraq, Poisonous Weapons, and Deterrence*, New Haven: Yale University Press, 1999, p. 237.
9 Hogendoorn, op. cit., p. 37.
10 Documentation online. Available HTTP:⟨http://www.cns.miis.edu/research/wmdme/libya.htm⟩ (accessed 5 October 2004).
11 For a description of UNSCOM's revelation of Saddam Hussein's Iraqi biological programme, see Tim Trevan, *Saddam's Secrets: The Hunt for Iraq's Hidden Weapons*, London: HarperCollins, 1999, particularly ch. 18.
12 Reuters, 'ElBaradei Urges Iran to Suspend Enrichment Activities'. Online. Available HTTP: ⟨http:// reuters.com/newsArticle.jhtml?type=topNews&storyID=6437647⟩ (accessed 7 October 2004).
13 Patrick Wintour, Brian Whitaker and David Teather, 'Blair Hails Libya Deal on Arms', *The Guardian*, 20 December 2003.
14 CNN, 2 December 2002.
15 Efraim Karsh, 'Israel's Imperative'. *The Washington Quarterly*, Summer 2000, pp. 155–161.

16 However, US–Israeli cooperation may be affected by US efforts to show the Arab world in the aftermath of the war on Iraq that the US is trying to balance its regional efforts.
17 Lisa Anderson, 'Absolutism and the Resilience of Monarchy in the Middle East', *Political Science Quarterly*, 106(1), 1991, pp. 1–16.
18 Mark N. Katz, 'Assessing the Political Stability of Oman', *Middle East Review of International Affairs*, 8(3) September 2004.
19 Eitan Barak, 'Where Do We Go from Here? Implementation of the Chemical Weapons Convention in the Middle East in the Post-Saddam Era', *Security Studies 13*, no. 1, autumn 2003, pp. 106–155.
20 In addition to these, only North Korea is probably a CW possessor state. Barak notes that the supportive evidence for including Myanmar seems to be 'quite far-fetched'. Ibid., p. 109.
21 Ibid.
22 Ibid., p. 112.
23 Barak has argued that the deadlock so far regarding Middle Eastern WMD disarmament – that the accession of Arab states to the CWC depends on Israeli access to the NPT – is based on three positions ('misperceptions'): that Israel has nothing to 'trade' regarding CW, that CW and NW are tradeable, and that linkage is a realistic policy. According to Barak, in principle these positions are changeable. Ibid., pp. 118–120.
24 Stephen M. Walt, *Revolution and War*, Ithaca, NY: Cornell University Press, 1996.
25 Marie I. Chevrier, 'Strengthening the International Arms Control Regime', in Raymond, A. Zilinskas (ed.), *Biological Warfare: Modern Offense and Defense*, Boulder, Colo.: Lynne Rienner, 2000, pp. 149–176.
26 Kenneth N. Waltz, *The Spread of Nuclear Weapons – More May Be Better*, Adelphi Papers No. 171, London: International Institute of Strategic Studies, 1981.
27 Hansen, op. cit.

8

MISSILE DEFENCE IN THE NORDIC COUNTRIES

Ingemar Dörfer

Introduction

This chapter will show why America and Europe are unequally interested in missile defence and the reasons behind this difference in opinion. The Nordic nations in particular are nuclear shy and have a great interest in arms control. Not being interested in the concept of nuclear deterrence, they have accepted the Cold War situation of strategic stability and now see any change in that strategy as a threat to stability. The missile threat is perceived differently in the United States and in the Nordic region. Thinking regionally rather than strategically, the Nordic nations are improving their extended air defence where cruise missiles form the greatest threat. However, even within the region perceptions and strategies differ.

The NATO membership of Norway and Denmark shapes their relationship to missile defence. They are engaged in the Western security debate, they have been used for many years to taking part in the implementation of NATO decisions, and they work with their allies, including the US, on many defence projects. Norway currently has two missile defence systems under development, the NASAMS II and the *Frithjof Nansen* frigate. Denmark is working on its extended air defence but is drawn into the American global missile defence system through its radar station at Thule. As former neutrals, Sweden and Finland have stayed outside the Western security debate. During the Cold War, Sweden took leading neutral positions on arms control and that is why the nation tried to take a stand on missile defence that collapsed when the ABM treaty was abandoned. Finland, having been more cautious owing to its former dependence on the Soviet Union, is concentrating on its territorial defence, including extended air defence.

The issue of missile defence was reinvented in America in 1998 fifteen years after President Reagan had introduced it in 1983.[1] While the Clinton administration procrastinated, it became a goal of the Bush administration with an initial deployment in July 2004. The main American argument for a missile defence has been to defend the homeland against rogue states, i.e. currently North Korea and Iran. But in addition there are strong reasons for creating a limited missile defence capability.

Since both American and European forces are expeditionary, theatre missile defence is needed to protect these forces against a future adversary with medium-range missile forces. Although the risk of an unauthorized launch is small, a limited missile defence would take care of the accidental stray attack. A credible American missile defence could also deter potential rogue states from developing intercontinental missiles with weapons of mass destruction, since the effort to overcome even a limited American defence seems overwhelming. Finally, in a crisis, American missile defence inhibits surprise attack by the enemy. Only a massive threat of dozens of nuclear weapons would be credible. Missile defence inhibits escalation.

The Europeans at first did not warm to the idea. The threat of a missile attack from a rogue state seemed small. Unlike Israel, Europe has not been subjected to Iraqi Scud attacks. Libya launched a missile on Lampedusa in the 1970s but the result hardly warranted an Italian missile defence system. Relations with Iran have mostly been friendly, and the thought of a North Korean missile attack on Europe is beyond the European imagination. The creation of European expeditionary forces presents a new challenge, but few Europeans believe that these forces will be sent into battle without American support. Theatre missile defence will then be provided by the US.

Another reason why missile defence in Europe is unpopular is of course the lack of funds. NATO planners know that the individual nations have great difficulties living up to the goals set by NATO summit meetings. The headline goals formulated for the Rapid Reaction Force of the European Union have so far remained on paper. A Europe that can barely live up to its formal defence commitments has no money for missile defence in addition. Thus, while NATO missile defence studies absorb $10 million a year, the American effort is one thousand times bigger – $10 billion.

Even if the money was allocated, the European defence technology is unable to handle the scale and sophistication of the American effort. Much duplication and waste would be the result. Better, then, to draw upon ready-made American systems, if and when a European effort is warranted.

The difference between America and Europe explored by Robert Kagan also applies to missile defence.[2] Where the Americans want hardware for war-fighting purposes, the Europeans prefer soft power to avoid war situations. Arms control and negotiations are better suited to solve the missile problem, as for all other problems observed by post-modern Europe. The case of Iran will show if this is a correct judgement. Thus the demise of the ABM treaty upsets the Europeans since it is seen as a blow to traditional arms control. Interestingly enough, the nation most concerned with the

ABM treaty, Russia, has not reacted according to the European stereotype. This is because Russia like America is not, according to Kagan, a post-modern society. Like America and unlike Europe it still has to conduct wars outside the European orbit.

Within Europe the Nordic region instinctively reacted the most to what is perceived as threats to the 'stability' of European security policy. During the Cold War the Nordic nations formed the only European nuclear-free space besides Switzerland/Austria and Yugoslavia. As a consequence very few politicians, except maybe in Norway and Denmark, mastered the concepts of nuclear deterrence and its implications.[3] Since a shift in nuclear strategy is the basis of missile defence, the shift is difficult to understand if you do not understand the original strategy to begin with. Old nuclear nations are less prone to see mutual assured destruction as 'stability'. Hence missile defence is seen through an arms control prism and the refusal to ratify the nuclear test ban treaty, together with the withdrawal from the ABM treaty, castigates the Bush administration in Nordic eyes as aggressive and disruptive in strategic matters.

Since the Nordic nations are split between NATO members and former neutrals, missile defence occupies a different place in the nations' political agendas. Because of NATO it appears as an issue that has to be dealt with in Norway and Denmark. In Sweden and Finland, on the other hand, interest in missile defence is an acquired taste to be cultivated according to one's own strategic preferences. Unlike nuclear weapons, it does not promote strong emotions and can by and large be kept out of the limelight.

When it comes to the Norwegian and Danish armed forces, their own force structures are subject to NATO guidelines and thus already include extended air defence in all planning. This is done through day-to-day work totally separated from any ideas on strategic instability or worldwide arms control.

But the military establishments of Sweden and Finland also carry on this work. Having concluded that missile defence is out of reach, politically and economically, they still study how the extended air defence of Sweden and Finland is to be shaped. Here, being outside NATO and without access to NATO programmes is at first no obstacle, since the main missile threat is cruise missiles. Access to American weapons systems is another matter. Without the F-18 Hornet and the 40 per cent American JAS with its AMRAAMs, the air defence would be dismal. Also Norway and Denmark are now concentrating now on the cruise missile threat, trusting that America will take care of any ballistic missiles.

The threat

In the CIA estimates of 1999 the main threat to America emanates from North Korea, Iran, Iraq, Syria and Libya.[4] Rogue states with long-range missiles and weapons of mass destruction are considered the main threat. In 2005 Iraq, Libya and to a large extent Syria have been eliminated. Of the remaining two, Iran has quite good relations with Europe while North Korea remains an enigma.

Since 2002 the Bush administration has made no distinction between national missile defence and a worldwide defence where the American national defence merges with regional and theatre defence. Thus the original European concerns of an isolated fortress America not willing to defend Europe, of decoupling, have mostly disappeared. To quote Henry Kissinger: 'A United States that is totally vulnerable to nuclear attack is much more likely to shrink from fulfilling Alliance obligations. These arguments apply as well to the defence of European territory against missile attack.'[5] None of the decoupling arguments have been heard from America's allies in Asia, a region that takes the missile threat much more seriously than Europe.

Since Europe so far has difficulties envisaging a threat from enemies armed with nuclear-tipped missiles, it no longer thinks about decoupling. Yet decoupling is exactly what al-Qaeda achieved in Spain by the use of terror, bringing the Spanish troops home from Iraq.

If the ballistic missile threat to Southern Europe seems distant, it seems even smaller to Scandinavia. There are few reasons why any rogue state would target Nordic territory. The most likely way Nordic troops would get into harm's way is instead through participation in out-of-area operations, with the new NATO Response Force in the case of Norway and Denmark, with the EU Rapid Reaction Force in Sweden and Finland. Whereas the NATO force will certainly have an American missile defence component, no such planning yet exists for the EU Rapid Reaction Force. In these forces the Scandinavians are dependent on others for their missiles defence. For their homeland defence it is cruise missiles that they see as a threat.

Cruise missiles

Cruise missiles, which came into vogue twenty years ago,[6] all but disappeared from the horizon for a long time (since only American ones were used) but have made a renaissance in the last few years. The foremost exponent of this threat is Denis M. Gormley. From an American perspective he states: 'Theatre-defence needs ought to take precedence over homeland ones, if only because it is in various regional settings that cruise missile threats are likely first to emerge overtly, and quite possibly in large numbers, to threaten US global presence.'[7] In December 2003 the Pentagon declared the cruise missile threat to the US to be exaggerated.[8] But in Scandinavia, theatre defence *is* homeland defence and the air defence systems of the nations have been organized accordingly. Despite the end of the Cold War it is Russian cruise missiles that occupy the Nordic military: the most in Finland, less so in Sweden, even less so in Norway and least of all in Denmark.[9]

In particular, two new missiles with an accuracy that enables them to inflict extensive harm with conventional warheads are now being introduced into the Russian forces: the Kh 55 Raduga with a range of 3,500 km and the Kh-101 Raduga with a range of 6,000 km. These missiles can be fired from the Bear, the Blackjack, the Backfire and the Flanker aircraft.[10]

Ballistic missiles

Thus, none of the rogue states is considered a threat to the Nordic region. Russia has been ruled out as a future aggressor in Norway, Denmark and Sweden, who are all transforming their forces from a territorial defence to military forces tailored for intervention abroad. Finland is also transforming its defence, but retains the territorial defence against the possibility of a Russian relapse into great power politics. To Finland, Russian ballistic missiles in the region are important.

According to the INF treaty, all Russian medium-range ballistic missiles have been scrapped. In the Leningrad military district, however, there were three missile brigades (Figure 8.1).[11] There were 2–4 nuclear warheads per missile and probably 100 altogether in the district. Two of the three brigades have been moved to reserve status and their materiel stored. The Luga brigade remains.[12] If it is converted into a SS-26 Iskander brigade its range will be 400 km, covering a large part of Finland. It is not covered by the INF treaty on missiles with a minimum range of 500 km. The Iskander is operational in 2004. It has a conventional warhead but can probably also use a nuclear one.

The NATO countries: Norway and Denmark

As NATO members, Norway and Denmark have to deal with the issue of missile defence because it shows up on the political agenda. NATO meetings are asked to take decisions on the question and NATO planners make plans and organize study groups and programmes where participation is mandatory. While missile defence is still on the backburner of such activity, even the NATO backburner adds many decisions and deadlines to the military-political bureaucracy of its member states.

Name	*Number*	*Range*
Pinozero SS-21	12	70–140 km
Kujvozi SS-21	12	70–140 km
Luga SS-21	12	70–140 km

Figure 8.1 Former missile brigades at Leningrad.

Issues such as nuclear weapons awoke strong emotional responses in the Nordic nations, with demonstrations and non-parliamentary activities influencing the decision-makers. During the battle of the Euro missiles in 1981 the Norwegian anti-nuclear opposition forced Norwegian Foreign Minister Frydenlund to bring up the question of a Nordic nuclear-free zone with Alexander Haig. Such a zone would lead to no US reinforcements being sent to Norway in wartime, answered the US Secretary of State.[13] The Norwegian activism then evaporated. Likewise Denmark during this period became a footnote nation, frequently taking issue with nuclear decisions at NATO meetings.[14] But the concept of a non-nuclear defence against ballistic missiles is too complicated for the public and too tame for the politicians to have any urgency. Arms control considerations are raised, but not as much as in Sweden. NATO membership means learning by doing also when it comes to missile defence.

Norway

In Denmark the Ministry of Foreign Affairs handles missile defence; in Norway, the Ministry of Defence. The two Nordic states bordering on the former Soviet Union, Norway and Finland, are the ones that still take the defence of that border very seriously. Norway is the nation that has the two most interesting weapon systems designated for missile defence. One is the NASAMS II ground-to-air missile and the other is the *Frithjof Nansen* frigate. Both are to be deployed first in 2006.

But Norway's participation in NATO deliberations and studies makes it the most sophisticated nation in the north when it comes to missile defence. Unlike the Netherlands with its Patriot PAC-2 missiles, missile defence is not a niche speciality to Norway, but its close cooperation with the Netherlands in defence matters has added to its knowledge of the subject.

In its transition to network-based defence the air defence is modular and mobile, able to move within Norway and out of area. As in other NATO nations, the extended air defence has four components:[15] (1) BMC4I, (2) conventional counter-force, (3) passive defence and (4) active defence. Norway supports the ongoing NATO/TMD/MD programmes through resources and personnel. A large number of NATO study groups are examining the various options.

At the Prague summit in November 2002 it was decided to 'examine options for protecting Alliance territory, forces and population centres – against the full range of missile threats'.[16] This should be achieved through an appropriate mix of political and defence efforts. A NATO feasibility study conducted by two industrial teams was won by the Science Applications International Corporation of Washington in August 2003 together with Boeing, EADS, QinetiQ and IABG of Munich, Diel Munitions and the Netherlands Organization for Applied Scientific Research. The follow-on study due in 2005 is to determine the feasibility of meeting the requirements as well as the associated performance, time scales, costs and risks.[17]

- To identify the range of eventual numbers of these systems this will be needed.
- To identify possible industrial strategies within which NATO might pursue the acquisition, in-service operation and support of TMD. A NATO capability by 2010 is the goal.

In its national programme Norway is focusing on the lower layer, terminal phase of missile defence. That means a joint forces approach and cruise missile defence capability.

NASAMS II

NASAMS II is a new version of NASAMS I developed by Kongsberg and Raytheon.[18] It is a ground-to-air version of AMRAAM batteries with a range of over 20 km. The first NASAMS II are to be deployed with the Norwegian Army Sixth Division in the far north in 2006. Spain has acquired a version for a further development and the Netherlands has also shown interest. NASAMS II can be moved around Norway and out of area to support other nations. It covers the low end of the anti-missile spectrum. The Netherlands with its future Patriot PAC-3 missiles is expected to provide the high-level defence for Norway and Denmark in a crisis. NASAMS II is commanded by the Norwegian Air Force. For its conventional air defence the Air Force has F-16 Fighting Falcons to be replaced by 2010. The main candidates are the US Joint Strike Fighter and the European Typhoon.

The Frithjof Nansen frigate

The only Scandinavian weapon system with a future ballistic missile defence potential is this Norwegian frigate. In 1993 the Storting decided to replace the increasingly obsolete Oslo-class frigates with six new ones. Developed and built in Spain, the *Frithjof Nansen* belongs to a new class of frigates in NATO with overlapping but not identical characteristics (Figure 8.2).[19]

The German, Dutch and Spanish frigates are built under the Trilateral Frigate Agreement. Norway is not part of this agreement but has drawn upon the technology developed by the other nations. Five ships are to be commissioned between 2006 and 2010. The main task is anti-submarine warfare and above-water warfare, protecting the Norwegian energy resources in the North Atlantic but also participating in allied out-of-area operations. The sensors and communications systems of the *Frithjof Nansen* are among the most sophisticated in the world, including the Aegis system,

Class	*Country*	*Number*
Sachsen-class F-124	Germany	3
LFC frigate	Netherlands	4
F-100 *Alvaro de Bazan*	Spain	4
Frithjof Nansen	Norway	5

Figure 8.2 New NATO frigates.

SPY 1-F radar as well as Link 16 and 22.[20] For its air defence it has Evolved Sea Sparrows to be launched by the MD 41 vertical launching system. But the MD 41 can also launch the US Navy Standard Missile-3 if necessary. The *Frithjof Nansen* is at 5,300 tons as sturdy as the *Alvaro de Bazan* at 5,800 tons. The Spanish frigate is the first non-US ship with the Aegis weapon system. In July 2003 it took part in combined Combat Systems Ship Qualification Trials with the US Aegis-class destroyer, USS *Mason*. The Spanish destroyer has the AN/SPY 1-D radar but will switch to the SPY 1-F radar in 2004, the same as the *Frithjof Nansen*. The Link 16 is crucial for communication with other ships in a seaborne missile defence group. If Norway in the future acquires the Standard Missile-3, the frigates can instantly participate in ballistic missile defence operations. Its theatre missile defence can cover substantial parts of Scandinavia.[21]

The Norwegian future

The defence study of the Norwegian defence chief stated in 2000:

> Missile technology combined with weapons of mass destruction in the hands of rogue states is a possible security threat in the future. Even if such a threat may concern Norway this study can see no possibility that we could establish a national defence in this area. We must realize that such defence systems could be developed jointly within the alliance over time. Norway should then consider participating in such cooperation when it is feasible.[22]

A current study carried out by the Norwegian Defence Research Establishment is considering various options for Norwegian missile defence. Norway can participate in a multinational missile defence in various ways:[23]

- Further improvement of the NASAMS.
- Developing the TMD capability on new frigates.
- Purchase/lease/pooling of Patriot/MEADS integrated with NASAMS.
- Financial support to NATO commonly funded TMD/MD programmes.

Kongsberg is the only Nordic corporation with a stake in missile defence. The Swedish defence industry has shown a total lack of interest, as has the Swedish government. Thus, missile defence and its national security implications are better known in Norway than in the other Nordic nations but only among a small military-industrial national security elite.

Denmark

In June 2003 the Danish government presented its study on new priorities in Danish foreign policy. It noted the emerging NATO cooperation on missile defence and also how Russia could get involved in the defence within the framework of the NATO–Russia council.[24] Like Norway, Denmark participates in the ongoing evaluation of missile defence. It sees the NATO Response Force as a key element in the transformation of Danish defence and it regrets that Denmark is prevented from participating in the European defence effort, notably the European Capabilities Action Plan.[25]

In September 2003 the Defence Ministry announced its conclusions based on a year-long study of Danish defence. The new Danish defence is to be transformed according to the RMA and is to go from a mobilized defence to a deployable network-based defence.[26] Since there is no specific missile defence capability, its defence against cruise missiles will, like the Norwegian one, be F-16 Fighting Falcon fighters with AMRAAM missiles. The only current candidate to replace the F-16 is the American Joint Strike Fighter. Unlike Norway, Denmark has no NASAMS missile and the Hawk, also deployed in Sweden, is to be phased out. Like Norway, Denmark is relying on allied help when it comes to missile defence, primarily the Dutch Patriot PAC-3 when the Netherlands acquires them.

Thule

From the beginning of the new American missile defence effort it was clear that the Thule installation would sooner or later be involved. The question of Thule was raised in Parliament and the Danish government published a report on the issues involved, which, like the British one, was quite basic.[27] On 17 December 2002, the same day President Bush announced a first deployment for November 2004, the formal request was given to the British and Danish governments to improve the radar facilities at Fylingdales and Thule. Whereas London agreed on 5 February 2003, the Danish–American negotiations lasted fifteen months longer. The Thule installation has, since 1961, been one of five ultra-high frequency early warning radars (UHF/EWRs) that provide both North American Aerospace Defense Command (NORAD), located at Cheyenne Mountain Air Station, Colorado, and the Unites States Strategic Command (STRATCOM) and others with early integrated threat warning and attack assessment of ICBMs and SLBMs penetrating the EWR coverage. Their missions also support the USSPACECOM Space Surveillance Network with Earth Satellite Vehicle surveillance, tracking and radar Space Object Identification. The other four EWR sites to be upgraded are located at Fylingdales, England, and Clear, Alaska, and the Cape Cod, Massachusetts, and Beale Air Force Base, California, Pave Pew radars.[28] The Fylingdales upgrade will be finished by late 2005, that at Thule by late 2007 at an investment of $260 million. They will cover missiles coming from the Middle East in contrast to the US radars in 2004 that cover North Korea.[29]

The Fylingdales radar also supports UK defence objectives. This is one difference between Fylingdales and Thule. British personnel are active at Fylingdales whereas the Thule radar is run by Americans only. Whereas Greenland has had home rule since 1979, the Fylingdales radar is in northern England, not Scotland. Scotland is on the way to home rule and had the American radar been based in Scotland according to the original American wishes in the 1950s, the negotiations would have been more complex. Britain has an emerging missile defence industry and been offered cooperation through which it could gain substantial contracts. Denmark has not, and has instead been offered other advantages such as contributions to the infrastructure and scholarships to Greenlanders.

Representatives of Greenland have been involved throughout the negotiations concluded in May 2004. On 26 May 2004 the Danish Parliament approved the new agreement. The agreement concerns only the upgrading of computers, software and communications of the early warning radar at Thule. It is currently estimated that an X-band radar at Thule may not be necessary.[30]

On 6 August 2004 the 1951 US–Danish defence agreement was amended with the upgrading of Thule. The signatories were the United States, Denmark and Greenland. This of course had been the only conceivable outcome all along. Greenland had been the main reason for including Denmark in NATO in 1949.[31] Denmark had been the only Nordic nation to support the war in Iraq, sending troops to help in the reconstruction. Domestic Danish opposition to missile defence had essentially disappeared with the demise of the ABM treaty in June 2002. The main debate in the Danish Parliament had taken place one year earlier in April 2003.[32] Here the traditional arms control arguments against missile defence had been ventilated. But there was no fire in the opposition, and the issue never caught on as a very important political question.

The former neutrals: Sweden and Finland

Throughout the Cold War, Finland and Sweden were neutral, Finland by necessity, Sweden by choice. Although limited by its Treaty of Friendship,

Cooperation and Mutual Assistance with the Soviet Union, Finland managed to maintain good relations with the US and its Western allies. In Sweden there developed a dichotomy between the neutral dogma preached by the Foreign Ministry and the excellent secret relations between the armed forces and the Pentagon and the British Ministry of Defence.

The end of the Cold War led to different reactions in the Swedish and Finnish leaderships. Sweden thought that the reunification of Germany and the disappearance of the Warsaw Pact eventually would lead to the dissolution of NATO as well and the creation of an all-European security system similar to OSCE.[33] Finland on the other hand did everything in its power to abandon its Cold War dependence on Russia. The old treaties with the Soviet Union were cancelled as well as restrictions on the size and shape of Finland's armed forces. While Sweden joined the European Union for economic reasons, the main Finnish argument for joining were the security reasons.

Because of its precarious position during the Cold War, Finland had never excelled in arms control initiatives, not wanting to antagonize important friends in the West. The Finnish proposal of a Baltic nuclear-free zone was a method of keeping Moscow happy. When the Swedes in 1983 seemed to take it seriously, Finnish diplomats cautioned Stockholm to cool it.[34]

This background is necessary to understand why Finland comes from a cautious minimalist policy in arms control while Sweden has pursued an activist arms control and disarmament policy. In 1989 the European security situation seemed to have stabilized, with arms control treaties regulating the nuclear relations between America and the Soviet Union and the CFE treaty the conventional military balance in Europe. Although Sweden was party to none of these treaties, it had signed the nuclear proliferation treaty and other agreements regulating chemical and biological weapons. It advocated nuclear disarmament in all its forms and was active in fighting the proliferation of weapons of mass destruction.

Like other Nordic politicians, the Swedish leadership shied away from all nuclear issues. Since the US nuclear deterrent covers Sweden as well as the NATO nations, there was no reason to talk about it or to pretend that it existed. A general feeling existed that the European/US/Russian arms control system was stable and that innovations caused a threat to that stability.

Stability of course was not the main purpose of the Bush administration in 2001 and its new nuclear strategy.[35] America was now vastly superior to Russia in power and resources and the old nuclear treaties from the 1970s reflected a situation of parity that no longer existed. The ABM treaty, which guaranteed mutual assured nuclear destruction with the USSR, also prevented the American build-up of a missile defence against rogue states. While Russia understood, Sweden did not, and in 2002 Sweden was the only

European nation except Russia that urged the US not to deploy a missile defence.[36] When Russia in June 2002 accepted the demise of the ABM treaty, European pleading on behalf of Russia lost its value.

But Sweden had also lost its old forum and not found a new one. The UN has not been interested in the missile defence issue and treated it marginally. The European Union has split here as on many other recent national security questions. Sweden has not managed to rally other Europeans to take a stand on the topic since most EU members now are also NATO members and since some of them, such as the UK, Poland, Hungary and Denmark, support the effort strongly.

Not being members of NATO, Sweden and Finland are limited to the defence policy deliberations of the European Union. In these talks, missile defence has taken a back seat compared to other defence issues. The Union is well aware of the deficiencies of the projected European Rapid Reaction Force. Already a defensive expeditionary force needs a theatre missile defence.[37] Lack of this component is pointed out in the White Paper on European defence published by the EU Institute for Security Studies in May 2004.[38] Acquiring such a force is urgent, states the study.[39] In reality, the Rapid Reaction Force is planned for deployment together with American forces. When and if it develops its own theatre missile defence system it will not be as a Swedish or Finnish component. The Dutch with American PAC-3 missiles and the Germans and Italians with MEADs are the main candidates for this task.

Sweden

Given Sweden's good credentials in fighting the proliferation of WMD, the government has prudently toned down its resistance to an American missile defence system that will be deployed anyhow. Meanwhile military planning for extended air defence is continuing.

In the air defence study of 2003, ballistic missile defence is not discussed. The main threat is a conventional threat from strike aircraft accompanied by attacks with cruise missiles. Air defence is supplied by JAS and Viggen interceptors. Armed with AMRAAM missiles, they are effective also against known cruise missiles.

A special feature is the RBS 23 Bamse ground-to-air missile developed by Saab Bofors Dynamics and Ericson Microwave Systems.[40] Ready since 1999, it has an altitude range of 12 km and a range of 15 km and can protect air bases, harbours and army units. No decision has yet been taken on production, but there are options for land-based Bamse as well as placing them on new corvettes. Sweden is contributing 2,000 men to the EU Rapid Reaction Force. The theatre missile defence for that force has not yet been determined. Given its niche capability, the Netherlands has been designated to study this element of the force.

Finland

The F-18 Hornet is the backbone of Finnish air defence. Together with AMRAAMs it is to defend Finnish air space against combat aircraft as well as cruise missiles. Helsinki is defended by the Russian SA-11 Gadfly air-to-ground missile. It has an altitude of 15 km and a range of 30 km. Finland keeps substantial territorial forces and has not rushed into network-based warfare in the way that Sweden, Norway and Denmark are attempting to. This signifies Finnish solutions unique for Finland that gives the defence a profile designated for a successful Finnish territorial defence against a potential future Russian enemy. Yet, like Sweden, Finland is studying the American Link 16 as a potential grid to coordinate future defence within NATO.[41]

Conclusions

As in the rest of Europe, the Scandinavian public and most of its politicians do not take missile defence seriously. As Robert Cooper, Javier Solana's chief foreign policy adviser, states, it will take a European 9/11 to make a difference in European attitudes.[42] Europe at this time has neither the money nor the technology to invest much in the project. Throughout the Cold War, Europe has been used to living under the nuclear threat. In America the Cuban missile crisis of 1962 made the public aware of the threat, but the rest of the time this was never an important political issue in American politics. It was out of sight and out of mind.

Missile defence does not have the emotional loading of nuclear weapons and the Scandinavians, like other Europeans, would rather have the Americans take care of it, like nuclear deterrence. When it does appear on the political agenda, as with the Thule radar, it is handled with a minimum of fuss, in a businesslike manner.

Among some Scandinavian leaders, formed by Cold War arms control thinking, missile defence is seen as a threat to 'stability'. Like other Europeans, the Scandinavians prefer diplomacy and negotiations to military force when solving potential conflicts. But as Sweden has noticed, things have changed. The United Nations is not the security forum it used to be. The European Union is deeply split on this as on many other security issues. America is more predominant than ever and other nations, especially Russia, go along, seeking cooperation rather than conflict. The example of Iran shows how difficult and dangerous the diplomacy route can be when it comes to nuclear proliferation.

In Europe, developing a theatre missile defence proceeds at a leisurely pace. 2010 is the goal for NATO, the initial operational capability of the US–German–Italian MEADS lies beyond the year 2015.[43] Other parts of the world are more worried. Israel's Arrow system was operational in 2004.

Japan is aiming for 2007. Australia is also getting involved. Most likely the bilateral systems developed by the US Navy and European navies will come into being earlier than the bureaucratic and underfunded NATO system. The Royal Navy, the Dutch, Spanish, German and Norwegian navies will be first. The Nordic militaries are planning and developing extended air defence systems as always. Defence against cruise missiles is one of the main tasks. Theatre missiles defence can be achieved only together with the Americans.

Notes

1 *Executive Summary of the Report of Committee to Assess the Ballistic Missile Threat to the United States*, Washington, DC, 15 July 1998. The Chairman was Donald Rumsfeld.
2 Robert Kagan, *Paradise and Power: America and Europe in the New World Order*, London: Atlantic Books, 2003.
3 The best analysis ever of Scandinavians and nuclear weapons is John Jörgen Holst, 'The Pattern of Nordic Security', *Daedalus*, Spring 1984.
4 Ingemar Dörfer, *Ballistic Missile Defence. Det amerikanska programmet. Säkerhetspolitiska konsekvenser*, Stockholm: FOI, April 2002, p. 5.
5 Henry Kissinger, *Does America Need a Foreign Policy?*, London: The Free Press, 2003, p. 68.
6 Richard K. Betts (ed.), *Cruise Missiles: Technology, Strategy, Politics*, Washington, DC: Brookings, 1981.
7 Dennis M. Gormley, *Dealing with the Threat of Cruise Missiles*, Adelphi Paper 339, London: IISS, 2001, p. 13; see also Dennis M. Gormley, 'Missile Defense: Myopia: Lessons form the Iraq War', *Survival,* Winter 2003–4; Dennis M. Gormley, 'The Emerging Cruise Missile Threat', *IQPC Missile Defence Conference*, London, 4–5 December 2003.
8 Dov Zakheim, 'No Need for a Cruise Missile Agency Right Now', *Inside Missile Defense*, 24 December 2003.
9 The Danish White Paper on the Thule radar lists the ballistic missiles of Iraq, Iran, Saudi Arabia, India, Pakistan, North Korea, Libya, Syria, Egypt and Israel but not Russia or China. Report from the Danish Government, *Missilforsvar og Thule-radaren. Redegørelse fra regeringen*, Copenhagen: Ministry of Foreign Affairs, March 2003, p. 11.
10 'Russia Deploys Conventional-Warhead Cruise Missile', *Jane's Military Review*, May 2003.
11 Ingemar Dörfer, 'Kola Has Lost Significance', *US Naval Institute Proceedings*, March 2002, p. 82.
12 Stefan Forss, *Venäjän ydinasevoimat 2004. Arvio niidennykytilasta ja tuelvaisuudesta*, Helsinki: VTT Energy, 2004.
13 Rolf Tamnes, *Oljealder 1965–1995*, Oslo: Universitetsforlaget, 1997, pp. 125–126.
14 Ingemar Dörfer, 'Scandinavia and NATO à la Carte', *The Washington Quarterly*, Winter 1986.
15 Tonje Skinnarland, 'Norwegian Views on Missile Defence', *IQPC Missile Defence Conference*, London, 4–5 December 2003.
16 NATO, *Prague Summit Declaration*, 21 November 2002.
17 Manfred Lange, 'NATO and Missile Defence', *IQPC Missile Defence Conference*, London, 4–5 December 2003.

18 *Jane's Land Based Air Defence, 2001–2002*, London: Jane's, 2001, pp. 290–291.
19 Online. Available HTTP: ⟨http://www.naval.tecknology.com/projects⟩.
20 Online. Available HTTP: ⟨http://www.mil.no/fregatter⟩.
21 I predicted this in 2002: *Svensk Tidskrift*, 6, 2002, p. 30.
22 Ministry of Defence, *Forsvarssjefens Forsvarsstudie 2000. Slutrapport*, Oslo: Ministry of Defence, 2000, p. 26.
23 Skinnarland, op. cit.
24 Ministry of Foreign Affairs, *En verden i forandring*, Copenhagen: Ministry of Foreign Affairs, 2003, p. 26.
25 Ibid., pp. 31–32.
26 Ministry of Defence, press release, 2 September 2003.
27 Ministry of Foreign Affairs, *Missilforsvar og Thule-radaren. Redegørelse fra regeringen*, Copenhagen: Ministry of Foreign Affairs, March 2003; Ministry of Defence, *Missile Defence: A Public Discussion Paper*, London: Ministry of Defence, December 2003.
28 *Upgraded Early Warning Radars (UEWR) for Missile Defense*, Bedford, Mass.: Raytheon Electronic Systems, 2002.
29 *Homeland Defense Watch*, 23 August 2004.
30 *Band Radar (XBR) for Missile Defence*, Bedford, Mass.: Raytheon Electronics Systems, 2002.
31 Don Cook, *Forging the Alliance*, London: Secker & Warburg, 1999, p. 193.
32 Folketinget, *Proceedings 2002-03-FB29-BEH1*, Tuesday 29 April 2003, Copenhagen. Since Russia had already accepted the end of the ABM treaty, the arms controllers debated the implications for China.
33 Henrik Liljegren, *Från Tallinn till Turkiet som svensk och diplomat*, Stockholm: Timbro, 2004, ch. 15.
34 René Nyberg, 'Security Dilemmas in Scandinavia: Evaporated Nuclear Options and Indigenous Conventional Capabilities', *Cooperation and Conflict*, 1, 1984.
35 Department of Defense, *Findings of the Nuclear Posture Review*, Washington, DC: Department of Defense, 9 January 2002.
36 Ingemar Dörfer, 'USA:s missilförsvar är inte långt borta', *Svensk Tidskrift*, 6, 2002.
37 Institute for Security Studies, *European Defence. Proposal for a White Paper. Report of an Independent Task Force*, Paris: Institute for Security Studies, May 2004, p. 119.
38 Ibid., p. 86.
39 Ibid., p. 122.
40 *Jane's Land Based Air Defence, 2001–2002*, London: Jane's, 2001 pp. 305–306.
41 The transatlantic 'grid' is a concept developed by David Gompert, Richard Kugler and Martin Libicki in *Mind the Gap*, Washington, DC: National Defense University, 1999, ch. 4.
42 'I believe Europe will have its own September 11', *Daily Telegraph*, 20 May 2003.
43 *Inside Missile Defense*, 7 July 2004.

9

NEGOTIATING BASE RIGHTS FOR MISSILE DEFENCE

The case of Thule Air Base in Greenland

Kristian Søby Kristensen

Introduction

For some time it has been American policy to try to include friends and allies in both the development and coverage of the missile defence system. The project is actually global, involving more and more countries, both as partners and as hosts for missile defence architecture.[1] Accordingly, during the development and deployment of the missile defence system the US has to engage a range of international actors. The engagement with these – friendly and often allied – actors is going to follow local dynamics and success is thus dependent on US management of regional politics.

The intersections between the global project and local dynamics have become even more important now that the Bush administration has abandoned the 'national' project of the Clinton administration in favour of a global design.[2] To go global means that the US must negotiate agreements with nations that potentially will host missile defence infrastructure and thereby must make missile defence a concern for allies as well. To investigate and answer how the US can handle the intricate links between local politics, allied relations and global defence issues is the subject matter of the following.

To help provide that answer, the chapter focuses on a particular case, Greenland. The island is part of the realm of Denmark but enjoys wide-ranging independence. Further, the island is host to Thule Air Base, an important node in the American early warning radar network. At present, after long negotiations, the US is free to upgrade the radar to be part of the missile defence radar architecture as well. The case is intriguing for several reasons. First of all, the US has already negotiated an agreement to upgrade the radar facilities, and we can thus track the process from beginning to end. Second, the process reveals how even old allies – Denmark was a founding member of NATO – can find it difficult to reach an agreement in a new era. Finally, it shows how local disagreements between actors in Denmark and Greenland complicate matters and thus influence the strategies of the actors.

Because of its wide-ranging autonomy and the contentious history of Thule Air Base, the indigenous population of Greenland plays a central role in the process, meaning that there are in fact three actors in the missile defence negotiations. The particular dynamics of the relationship between the actors entail that the process associated with the American request can fruitfully be analysed as two separate negotiations. First a primarily internal negotiation in the Danish Realm between the Danish government and the Greenlandic Home Rule Government took place. This negotiation was a precondition for formulating joint Danish–Greenlandic demands for the next round of negotiations, this time mainly between the US and Greenland. The central role of Greenland is reflected in the structure of the negotiations which are primarily concerned with defining a new role for the Greenlandic–American relationship, in essence meaning increased Greenlandic influence on the US defence areas in return for agreement on the missile defence upgrade. This process was long and hard, taking more than a year and a half, and straining the relationship of the actors. In the end, a workable solution was reached, satisfying all three parties. However, at the same time all three actors had to make concessions to reach the final agreement, and only by applying rather hard pressure on the Greenlandic negotiators was the US finally able to secure its right to upgrade the radar at Thule.

The chapter accordingly proceeds by elaborating the historical and political background to the negotiations. Special attention will be given to the historical role of Thule Air Base and the consequences of its history. This paves the way for the analysis of the actual negotiations concerning the American request to upgrade the facilities. The actual analysis is split into two sections. First, the more or less domestic debate between Denmark and Greenland is investigated. This is then followed by an account of the second round of negotiations between Greenland and the US. Finally, a conclusion will summarize and discuss the findings of the case and put the lessons learned about this engagement in perspective in relation to the further globalization of the US missile defence.

The controversial history of Thule and its implications for the missile defence negotiations

The American air base at Thule in northern Greenland was established in 1951 at the very beginning of the Cold War. Its costly establishment was in itself an expression of the increased importance of the Arctic region in the beginning confrontation between the two major Cold War powers. The role of the air base has evolved in tandem with the military and technological developments of the Cold War. Initially the base played an important role as a staging base for the bomber fleet of Strategic Air Command. Later, when ICBMs became the preferred means of delivery of nuclear weapons for both the US and the USSR, Thule Air Base became an important early warning radar site, tasked with detecting and tracking enemy missiles as well as space surveillance – a role it still plays today.[3] It would basically play the same role when upgraded and integrated in the architecture of the missile defence system.[4] Accordingly, the actual changes to the installations will be rather diminutive, mostly comprising software changes.

Thule Air Base, as well as the Arctic in general, has been important to the United States.[5] However, the base and the American presence on the island have probably been even more important to Denmark. The geopolitical position of Greenland, squarely between the two superpowers, was undoubtedly an asset for Denmark in negotiating NATO membership, and the value of Greenland has repeatedly shown itself in the alliance politics of the Cold War.[6] By playing the 'Greenlandic card', i.e. by allowing US military activities on the island, Denmark has been able to achieve concessions from the US in other areas.[7]

However, the status of Greenland within the Danish Realm changed during the Cold War as well, and several developments and incidents in time questioned both the authority and legitimacy of the Danish government. By the time of the establishment of Thule Air Base in 1951, Greenland was still a colony. Yet already in 1953 this changed. Greenland attained the status of a county and was thus integrated on equal terms in Denmark proper. This set the stage for the emergence of a political Inuit consciousness on the island and, with that, the wish for wider independence from Copenhagen. In 1979 this resulted in the passing of the Home Rule Act, delegating political authority on substantial policy areas to the locally elected Home Rule Government. One exception was that of foreign policy. Still, even in this area, a working relationship evolved; the Home Rule Government participates in most foreign policy decisions concerning Greenland, and in some cases makes its own decisions.[8] The Home Rule Government thus has to be taken into account by the Danish government, if its policy decisions concerning Greenland are to have any legitimacy. Because of the Cold War history associated with Thule Air Base, this is especially the case in relation to security policy and thus missile defence. A number of incidents related to Thule have played an important symbolic role in structuring the Danish–Greenlandic political relationship.

The first and most important incident took place in 1953. As a consequence of an enlargement of the base areas at Thule, a small community of Inuit was, against their will, moved from their traditional settlement near the base by the Danish colonial authorities. No one paid much attention to the issue at the time. It was handled quietly by the office of the Prime Minister,[9] and the public was given the impression that the reallocation was voluntary. It was not until the mid-1980s that the true circumstances were revealed.[10] The result was staunch criticism of the morality of the Danish government both in Denmark and in Greenland, as well as legal action against the Danish state questioning the legality of the reallocation.[11]

Another pivotal incident took place in 1968 when an American B-52 bomber carrying nuclear weapons crashed near Thule. This incident questioned the clear official Danish policy on nuclear weapons. Denmark categorically opposed the presence of such weapons on Danish territory, a policy that fitted badly with nuclear-armed B-52s flying through Danish airspace. This further made it clear that the base indeed posed a potential risk to the Greenlanders, both as a consequence of accidents and in case of a potential superpower conflict.[12] The question of nuclear weapons became a subject again in 1995, when a document appeared which showed that the Danish government in effect conducted two nuclear weapons policies.[13] This 'double play'[14] meant that the government publicly stated that there were no nuclear weapons on Danish territory, while since 1957 it had secretly been allowing the US to station nuclear weapons at Thule. The Danish government had played the Greenlandic card without the knowledge or consent of either the Greenlanders or the Danish Parliament.

This history shows Greenland that it is hard to trust the benevolence and truthfulness of the authorities in Copenhagen. The feeling in Greenland is that the Cold War history shows that the Danish government cannot always be trusted to act in the best interest of the Greenlanders. Put bluntly, their rights have been compromised by Danish governments' efforts to please the Americans. This history has for a long time made Thule Air Base an infected issue in the Danish–Greenlandic relationship and the history is still very important for both the Greenlandic self-perception and their perception of Danish actions. The historical background accordingly becomes an important element in the debates and negotiations concerning Thule Air Base and missile defence. Thus, after briefly clarifying the history, we can now turn to the present.

The upgrade debate in the Danish Realm: the first round of negotiations

Anticipating the possible role of Thule Air Base in the American missile defence plans, the issue of missile defence has been on the Danish–Greenlandic political agenda since 1999. The salience of the issue was by no means lessened when, on 18 December 2002 – just after President Bush announced his decision to deploy BMD in 2004 – US Secretary of State

Colin Powell made a formal request to the Danish government for acceptance of the American upgrading of the installations at Thule.[15]

At first it looked as if the Danish government would easily reach a positive decision. It appeared to be in a parallel situation to the British government, which already on 5 February 2003 was able to respond positively to the American request to upgrade the identical radar at Fylingdales.[16] That was, however, not the case with Thule. Not before 26 May 2004 was the Danish Foreign Minister able to announce that an agreement concerning the upgrade of the radar was in place, only six months before the missile defence system was to be declared operational.[17]

The Danish government did not hide the fact that its preferred policy was to follow that of its British colleagues and provide its consent as quickly as possible. Nonetheless, more than a year and a half went by before a decision was reached. Basically the process dragged on because the Home Rule Government had a legitimate claim to be included, and its concerns to some extent had to be addressed by the other two parties.

The public process of discussing the conditions for upgrading the radar at Thule was thus kicked into motion by the request made by Secretary Powell. However, the script for the negotiations had for some time been prepared by the three actors, and their initial positions had more or less been stated. The request itself came as no surprise, and the time of its publication was probably to some extent coordinated between the three actors.[18] The initial positions on missile defence and how to reach a decision on the request were thus relatively clear. The US had, both previously and at the time of the request, made clear that it was conscious of the domestic implications of its request, and appreciated the need for an internal debate in the Danish Realm before any reply could be made.[19] Equally, the Danish government had stated that in general its attitude to the American plans was positive, but with the qualification that any decision should be preceded by a thorough democratic debate, and in cooperation with the Home Rule Government.[20] Finally, the Home Rule Government demanded to be incorporated in the process, thus increasing its influence on security policy and especially a renegotiation of the old agreement regulating the US defence areas. Further, a demand was made for payment of rent directly to the Home Rule Government.[21]

Accordingly, the stage was set for the first round of negotiations. The US in effect withdrew, stating that it would be improper to participate in what was labelled as an internal process in the Danish Realm, thus effectively putting the responsibility for reaching a compromise on the shoulders of the Danish government.[22] The Danish government and the Home Rule Government then had to agree on joint conditions for allowing the upgrade before the US would involve itself. As we shall see, this put pressure on Denmark, which had to reach an agreement with Greenland while simultaneously taking the wishes of the US into account.

Constraining the Danish government

This marks the beginning in earnest of the first round of negotiations. Denmark and Greenland had to reach a decision before taking the issue up with the US. These negotiations were, however, not easy. The Danish government had put itself in an awkward position where potentially conflicting demands were being placed on it. First, without being outspoken, the US nonetheless put pressure on the Danish government to reach a (positive) decision as soon as possible, and by disengaging itself the US effectively put the responsibility for reaching that decision on the shoulders of the Danish government. Second, the Danish government had to find a compromise satisfying Greenlandic demands while at the same time being acceptable to both itself and the US. To complicate matters further, the Danish government had declared that a result would be reached only through a thorough and informed democratic (meaning public) debate.[23] The Danish government was thus caught in a dilemma: a decision on one hand had to be reached as quickly as possible but with adequate time to conduct deliberative democratic discussions, and on the other hand the decision had to be acceptable to the central ally, the US, as well as to Greenland without whom any decision would lack legitimacy.

These potentially conflicting considerations created ample room for the Greenlanders to engage the Danish government and to pursue their own demands against an opponent on the defensive. The very fact that the role of Thule Air Base was now, in connection with the missile defence debate, on the political agenda provided the Greenlanders with an opportunity to expand their autonomy and increase their influence on security and defence policies. The most effective strategy for doing this was to use the history to practise the politics of embarrassment.[24] It is not only in Greenland that the Cold War history of Thule Air Base was considered morally problematic. It does not fit the way Denmark likes to picture itself, nor does it fit the assumptions guiding present Danish policy towards Greenland.[25] By publicly referring to how Denmark acted towards its indigenous population during the Cold War, in a way that is currently viewed by all parties as both embarrassing and unjust, and then equating the past with the present, the Greenlanders forcefully strengthened their position in the negotiations with the Danish government. By using the history to practise the politics of embarrassment, the Greenlanders attempted to use their moral authority to constrain the actions legitimately taken by the Danish government in the negotiations concerning the upgrade of Thule Air Base.[26] To accomplish this, the Greenlanders tried to turn the missile defence debate away from missile defence proper and instead to focus the debate on the democratic character of the debate itself.

An important event – both in the debate itself and for the effectiveness of the Greenlandic strategy – was the publication of the Danish government's

White Paper, *Missile Defence and the Thule Radar*, intended to stimulate the democratic debate promised by the government.[27] The stated purpose of the White Paper was to present the issue, 'to make a complicated issue comprehensible and to cast light on many of the questions that rightfully can be raised concerning missile defence'.[28] Missile defence, as presented by the government in the White Paper, in essence poses no negative consequences, either narrowly for Denmark or internationally. The White Paper in large part follows the American arguments for missile defence. It states that both WMD and missile technology are indeed proliferating into the hands of actors against whom traditional arms control diplomacy and deterrence are not viable political strategies – the magnitude of the threat is increasing – and a possible countermeasure is missile defence.[29] Missile defence is thus a purely defensive system. The threat stems from irrational rogue states, with only a few missiles at their disposal, thus not threatening the inhabitants near Thule Air Base. The actual architecture of the system will of course reflect the size of the threat, and the system will not pose a threat to the strategic stability in the relationship between the major powers still guaranteed by MAD.[30] The threat is, however, not exclusively directed towards the US, and as missile defence in time is supposed to cover 'friends and allies' as well, it is, according to the White Paper, in Denmark's national interest to facilitate the construction of the defence system.[31] Missile defence therefore makes sense from both the American and Danish points of view. The White Paper essentially draws a very positive picture of missile defence, in line with the government position.

The arguments of the White Paper were, however, not allowed to stand unopposed. On a number of accounts, commentators and the opposition questioned the analysis presented by the government. Whether the information and the arguments of the government as presented in the White Paper and elsewhere were true became a central issue in the public debate. Accordingly, the reception of the White Paper functions as a first-rate example of how the Greenlandic strategy worked in undermining the legitimacy of the Danish government by questioning its democratic intentions and truthfulness.

The first problem raised in the debate about the White Paper is, as the final configuration of the system is not yet known, the question of its ultimate role. A comprehensive multilayered system might actually undermine global strategic stability and potentially start an arms race.[32] Second, a functional missile defence might have offensive potentials, as it would reduce the risk associated with eventual American interventions abroad.[33] Missile defence thus cannot a priori, as stated by the Danish Prime Minister, be 'a Project for Peace'.[34] Finally, as missile defence might upset the strategic balance, and as the future threats are not known today, it cannot with certainty be argued that Thule will not again become a target, as it was during the Cold War.[35] The debate about the White

Paper opened up room for questioning the truthfulness of the Danish government. For example, by drawing on the criticisms above, it was possible for a Greenlandic member of the Danish Parliament to label the White Paper as 'feeble propaganda . . . without documentation, but filled with assertions'.[36]

Turning the debate from the issue of missile defence proper to the issue of honesty in the Danish administration and government in the debate itself, served the Greenlanders well. It questioned whether the Danish government truly intended to reach a decision in cooperation with the Home Rule Government on the basis of a genuine democratic debate. The Greenlanders came to stand out as being concerned with truth, peace and democracy, whereas the Danish government is presented as immoral and only seeking to please the US by securing an affirmative answer as soon as possible. By presenting the actions of the present Danish government as comparable with the actions of past Danish governments, the Greenlanders reinforced their moral superiority and thus their position in the negotiations. This is the case even though the Danish government at every opportunity stressed that any decision would be reached in accordance with the Home Rule Government and only after a democratic debate.[37] The success of this Greenlandic strategy thus of course in large part depended on convincing the public that the Greenlandic presentation and claims are valid. Both the opposition in the Danish Parliament and large parts of the media follow the Greenlandic agenda, and recount the history, making this the reason for legitimizing the Greenlandic demands against the government. For parts of the opposition a precondition for agreeing to the upgrade in fact becomes Greenlandic consent. Thus the unjust past is in the public sphere seen as central, and the government should act accordingly; alleviating the just Greenlandic demands in a democratic fashion becomes a demand placed on the government from all sides. It thus had difficulties acting in other ways.[38] On the basis of this public debate, the central issues of the negotiations thus became the question of what 'in accordance' actually means and thus the actual extent of Greenlandic influence on Danish security policy.

Concluding the first round

It is, however, not only the Danish government that is constrained by this public debate. The Greenlandic strategy obviously also affected which demands they themselves could legitimately put forward, depending on how well they fit the political strategy of embarrassment. The Greenlanders put forward three concrete demands: direct financial compensation; a renegotiation of the original agreement between the US and Denmark regulating the defence areas in Greenland, dating back to 1951 when

Greenland was still a colony; and finally, Greenland wished to be formally guaranteed participation and influence in general on security and defence policy issues of importance for Greenland.[39]

All three demands sought to be satisfied by referring to past injustices, but some fitted this strategy better than others. The Danish government easily dismissed the demand for rent as Greenland already received substantial financial support from Denmark. And as we shall see, the issue is not to be solved by the Danish government.[40] It was difficult to picture the lack of financial payments as essential to rectifying past injustices, and to make sure they will not happen again. The contrary was, however, the case with the so-called '1951 Agreement'. This old agreement embodies, for the Greenlanders, all that is wrong with the present situation. It dates back to when Greenland was still a colony; it makes little reference to the inhabitants of the island and gives them no rights in relation to the defence areas, either towards Denmark or the US.[41] Past injustice was perpetrated under the regulations of the 1951 Agreement, and, as argued by the Greenlanders, because it is still standing nothing has really changed. Thus a renegotiation of the agreement was of paramount importance. Accordingly, the demand was only partially dismissed by the Danish government, which argued that a genuine renegotiation of the agreement would have to be ratified by the American Congress. The Danish government essentially stated that it was beyond its capacity to deliver.[42] Instead the Danish government complied with the last Greenlandic demand, to be formally included in decisions on security and defence policy when they are of importance for Greenland.

Thus on 14 May 2003 the Danish Foreign Minister and the Greenlandic Home Rule Premier were able to sign a Declaration of Principle, wherein Denmark essentially agrees to the Greenlandic demands. The declaration stipulates that in issues of foreign and security policy of importance to Greenland, 'it is considered natural that Greenland be co-involved and maintain a contributory influence'. In addition, the declaration states that Greenland can be co-signatory with the Danish government with binding effect in international negotiations.[43] The declaration thus guarantees Greenland influence on, and participation in, security policy concerning Greenland.

Furthermore, a statement by both parties proclaims that the two sides agree on a joint proposal to be negotiated with the US including the 'renewal of the Defence Treaty of 1951' and the signing of an 'agreement on economic and technical cooperation intended to meet Greenland's wishes to increase its relations with the US'.[44] The statement thus opens a path whereby the Greenlanders themselves can negotiate an 'agreement on economic and technical cooperation', in effect meaning that it is up to the Greenlanders to negotiate bilaterally with the US any financial side payments as well as alleviating the Greenlandic problems with the 1951 Agreement.

In sum, the Greenlanders obtained important results. By achieving the main demands – increased participation and influence – the Greenlanders made sure that decisions cannot in the future be taken independently by the central authorities in Copenhagen. Second, the value of the newly gained right to participate will show immediately, as the fulfilment of the other demands will be determined through more or less bilateral negotiations with the Americans, the subject of the next section.

But what about the Danish government? It seems as if it gave in on most of the issues without getting anything in return. That is, however, not the case. If one looks at the settings for the negotiations, i.e. the dilemma noted earlier, and takes the confrontational Greenlandic strategy into account, the Danish government actually achieved substantial political objectives. Basically the fact that an agreement was reached can be said to be a victory for the government. It knew beforehand that the process would be difficult given the history of the Thule Air Base; it also knew that history made it imperative that the government did not act in any way that could be accused of being pushy, arrogant or undemocratic. Accordingly, the imperative for the government was to find a solution that could work for all parties, and maybe finally close the historical wound associated with Thule, or as formulated by a Greenlandic parliamentarian, to finally 'kill an old ghost from the past'.[45] This would effectively improve the future basis for Danish–Greenlandic relations, and reduce the future effectiveness of the politics of embarrassment. Further, being able to deliver a road map for reaching an agreement would obviously improve the already close Danish–American relationship, in itself an important goal for the Danish government.

Second, the nature of the agreements means that the direct role of the Danish government diminishes. The Danish government must have had some indication that the joint negotiation proposal to be discussed between the US and Greenland–Denmark would probably contain achievable goals and that the government would therefore be able to take a back seat. This was to be an opportunity for the Greenlanders to use their new right to negotiate with other states, and Danish involvement would be seen as undue interference with what was to be bilateral negotiations. The Danish government considered its task done.

The Danish government had effectively balanced domestic policy imperatives with mounting outside pressure. It delivered what was in its power to the Greenlanders, while not promising any more than what was believed the US would give in return for the Thule upgrade. Accordingly the Danish government was able to deliver to its big ally as well; it set the conditions for a bilateral Greenlandic–US agreement, which hopefully in the end would facilitate the American upgrade of the Thule radar.

Negotiating with the Americans: the second round of negotiations

This marks the active return of the US, which until now had been on the sideline, not overtly interfering and only awaiting the conclusion of the internal process in the Danish Realm. The following will analyse the second round of negotiations, this time between Greenland and the US. The first round of negotiations between Denmark and Greenland already defined the issues of the negotiations. What were at stake was a way to modernize the 1951 Agreement and the topic of compensation, in the form of actual financial rent or some other form of accommodation. The first issue – the modernization of the 1951 Agreement – caused no problems. The Greenlandic demand to be more actively involved in the American defence areas and to be included in some way in the 1951 Agreement was seen as both just and legitimate by the American negotiators.[46] Accordingly, both parties rather early on agreed on how to ensure this without an actual renegotiation. However, the second issue caused unforeseen problems. In light of the previous agreement and the fact that the US undoubtedly already was in some form of contact with the parties concerning this issue, the second round could have been expected to be short, almost a formality. As it was, the negotiations dragged on for almost a year, frustrating both parties.

As this is the first time in the process that the US is directly involved, the analysis now proceeds with a detailed discussion of the overall US strategy towards Denmark and Greenland in the missile defence negotiations. The understanding of the US strategy is necessary as it provides large parts of the explanation of why the process dragged on for so long. Arguably the Greenlandic participation also had an effect, and it will be integrated along the way. This will then be followed by a discussion of the final compromise.

The American strategy

The formal request on 18 December 2002 was, as stated earlier, not only directed at Denmark. The UK was questioned on the facilities at Fylingdales at the same time. Additional consultations were being undertaken with Canada, concerning the possible future of NORAD at a time when the missile defence would be operational and probably integrated into that system. All three countries have at times been lukewarm to the concept of missile defence. However, it is interesting to note that the US strategy for dealing with Canada and the UK on the one hand, and Denmark, including Greenland, on the other hand, on the issue of missile defence to a large extent differs.[47]

In dealing with the Canadians and the British, the US strategy has in short been to apply at times rather outspoken political pressure,[48] combined with promises of lucrative contracts for the defence industry in the two countries, as well as other political benefits, in return for their broad cooperation on missile defence and the right to use and station missile defence material on their soil.[49] Indeed Britain, and to a lesser extend Canada, have complied with the American wishes. However, both backbench and opposition criticisms to the American plans in both countries have been persistent throughout the process. And even though the prospect of defence industry cooperation has gained the US a domestic UK–Canadian constituency,[50] the American pressure has backfired at times when it has been seen as undue meddling with national political decisions.[51]

The above strategy is very different from the one that can be deduced from the case of Greenland. Even though a number of similarities exist between the countries, which, for instance, all are very close US allies, and all hold equipment important for missile defence on their territory, the differences are important. First, Denmark is a small country, and its outspoken support for missile defence is not as important to the US as that of the UK. Second, Denmark has no significant defence industry, and thus support cannot be gained by hinting at lucrative missile defence contracts to national industry. Third, the US was probably aware of the potentially powerful and outspoken domestic opposition over Greenland.

Whether the US strategy in the case of Greenland differs because of its assessment of this different situation, or because it was developed in cooperation with Denmark, or because it is a result of a 'learning process' indicating that an aggressive public strategy can backfire, or is a mixture of all three, is hard to say. What matters is the fact that the US engages Denmark and Greenland through what can be labelled as a strategy of non-involvement, and subsequently to analyse the consequences of that strategy.

Both before and after the official request was made, the US role has been very low key. The US limits its direct and public involvement to participating in Danish hearings, explaining the subject and restating the claim that it understands the need for a democratic debate and the need to integrate the Greenlandic Home Rule Government in the Danish decision, and that this in fact takes time.[52] This works to the benefit of the US. First of all because the US insulates itself from the politics of embarrassment. The US follows lines of action that in essence amount to respecting the moral authority of the Greenlanders by acknowledging the need to find a democratic solution that alleviates the Greenlandic worries and historical hardship associated with the air base at Thule. Seen in this way, nobody can justifiably question the morality of the US in the process. Second, the strategy of non-involvement lessens the US responsibility. The issue becomes a domestic Danish one, which has to be solved by the Danish government. As it would be inconceivable for the Danish government to refuse the upgrade, this in itself, in fact, makes the issue of finding a compromise within the Danish Realm a joint American–Danish strategic problem. By making the issue domestic, however, the responsibility for solving the problem is placed solely on the shoulders of the Danish government.

The American actions thus actually reinforce each other; by acknowledging the Greenlandic demands the US has to disengage, leaving the issue to the internal debate, and exactly by disengaging the US cannot be accused of acting undemocratically as was the case with the Danish government. Accordingly, the first round of negotiations make the US seem benevolent, making only positive statements, and acknowledging the necessity of a democratic debate as well as the legitimacy of finding a way to rectify the problems associated with the 1951 Agreement. However, this strategy of non-involvement at the same time entails that the initiative is handed over to the other two actors. This subsequently means a loss of control on the part of the US, the consequence of which shows itself in the second round of negotiations. What was to be a formality dragged on – partly as a result of the American strategy.

By disengaging, the US effectively sidelined itself. This in turn meant that the actors were out of synch with the particularities of the issues, and thus the US in reality lost control over the negotiations, inheriting an already solidified agenda in the second round of negotiations – an agenda not in the interest of the US, as it contains the difficult issue of rent for the base area at Thule, an issue that could set precedents worldwide,[53] making it impossible for the US to give up any ground. Neither could the Greenlanders in fact. The moral arguments of the Greenlandic politicians are of course not only heard in Copenhagen. The domestic Greenlandic opinion takes account of these arguments as well. It is exactly the moral character of these arguments which made it difficult for the Danish government not to concede to the Greenlandic demands, which now in turn limits the room of manoeuvre for the Greenlandic negotiators towards their American counterparts. Even the issue of rent takes on a moral character and is easily constructed as an absolute demand itself, as it potentially helps reduce the Greenlandic economic dependence on Denmark.[54] Furthermore the function of the rent issue for the Greenlanders was to show the tangible results of their new rights to bilateral negotiations to their domestic constituency.[55] Thus, during the actual negotiations when it became clear that this demand was not necessarily fulfilled, it was equally difficult for the Greenlandic politicians to back down and formulate new counter-demands.

The parties accordingly had to find some sort of compensation not involving actual payment but still satisfying the Greenlanders by making the value of the newly won right to negotiate bilaterally clear to the Greenlandic constituency. This was, as the agenda for the negotiations had solidified, not easy. Accordingly the negotiations went into deadlock. The deadlock of course meant that the American time schedule for the negotiations was sliding, and now the strategy of non-involvement worked negatively. It was difficult for the US to put hard and direct public pressure on the other actors when exactly the opposite had been the explicitly stated and morally argued line of action earlier in the process. Additionally, to apply public pressure would obviously still run the risk of being countered by various Greenlandic moral arguments. Furthermore, no help could be expected from the Danish government. By framing the process as two independent bilateral negotiations it would be just as morally untenable if the Danish government made its presence felt on the Greenlandic negotiators now as it would have been if the US had interfered in the first round of negotiations.

Accordingly, the US somehow had to solve the deadlock by itself, in direct negotiations with the Greenlanders. The US had to put pressure on the Greenlanders in the negotiations. Two strategies were employed: direct pressure in the negotiations combined with indirect public pressure. Even though no public statements were made by US officials concerning the lack of progress in the negotiations, it nonetheless became clear to the Danish–Greenlandic public during late winter and early spring 2004 that all was not going according to plan and American frustration was rapidly mounting over what was indeed considered an important issue in Washington. Both Colin Powell as well as an Assistant Secretary of Defense visited Copenhagen during winter and spring 2004, and while still downplaying the American frustration, the reason for these visits was not lost on the Danish media.[56] Thus without directly mentioning the problems in the negotiations, the clear impression conveyed by the American actions was nonetheless that the deadlock in the negotiations was increasingly considered a problem in Washington. This indirect pressure put the deadlocked negotiations on the public agenda. Further, the very fact that the schedule was sliding in itself became an important means to up the ante in the negotiations. It became publicly clear that, from a US perspective, the process was reaching the point of no return. Basically, US budgetary requirements related to the fast approaching deployment date were used as arguments to pressure the Greenlanders.[57] If a compromise could not be reached shortly, the US would be forced to seek alternatives making the base redundant.[58] This essentially amounted to an explicit threat to leave Greenland.

In sum, the process of – indirectly but effectively – making the negotiations public in effect put pressure on the Greenlandic negotiators to comply with the American demands. This strong pressure was combined with the promise of a wide-ranging framework for cooperation between Greenland and the US on a broad range of issues. This framework first holds the promise of contributing to the development of the Greenlandic society and infrastructure – but without direct financial payment. Second, this cooperation between Greenland and the US, if successful, could in time reduce Greenlandic dependence on Denmark. These two aspects of the framework give the Greenlanders something to show their domestic constituency. This was how in the end an agreement was reached. In essence, the

US used both stick and carrot. By making the future costs of non-agreement clear, setting a final deadline and showing the importance of the issue in the eyes of the American administration, while simultaneously presenting a solution with the future potential of satisfying important Greenlandic wishes, the US finally obtained compliance from the Greenlanders.

The conclusion must be that the US strategy was effective. The cooperative approach of non-involvement evidently helped the US to avoid being the target of domestic opposition in Denmark and Greenland. Simultaneously, it put large parts of the responsibility for reaching an agreement on the shoulders of the Danish government. However, at the same time the non-involvement strategy most probably contributed in part to the misunderstandings responsible for setting the difficult agenda in the second round of negotiations. This in turn – together with the difficulties faced by the Greenlanders – complicated the process and made it drag on, in the end causing grievances in Washington.

The final compromise

The prolonged process meant that not until 26 May 2004 could Denmark, Greenland and the United States present two joint declarations and an amendment to the 1951 Agreement to be signed later in the year, finally allowing the American missile defence upgrade more than a year and a half after the original request was made.

The agreements actually express a fundamental change in Greenland's relationship with the US and in its position within the Danish Realm. The most explicit fact picturing this change is that the Home Rule Government figures as a co-signatory of the amendment.[59] In addition to including the Home Rule Government as a party to the agreement, it is acknowledged in the amendment that Greenland's contribution to the mutual security of the parties should entail further Greenlandic rights and influence on the US base areas.[60] Therefore, the amendment stipulates that a representative of the Greenlandic authorities may be appointed to Thule Air Base, and that the US will inform and consult with the Home Rule Government before 'implementation of significant changes to the US military operations or facilities in Greenland'.[61]

The Greenlanders can rightfully be pleased with the amendment. In effect it guarantees that a re-enactment of the past is no longer possible. The Home Rule Government is through its representative closer to the actual activities at the base, it is guaranteed the same information as the Danish government, and finally, by being co-signatory to the legally binding amendment these rights are secured, as future amendments are possible only by mutual agreement by all parties. The amendment greatly changes the role of Greenland in the Danish Realm. In matters of security policy Danish sovereignty has essentially been shared with the Home Rule Government.

On the second issue, actual compensation – the Gordian knot of the second round of negotiations – a compromise was reached as well. By signing a 'Joint Declaration on Economic and Technical Cooperation' both parties were satisfied.[62] The US avoided setting perilous precedents of direct payment for other bases while at the same time rewarding Greenland for allowing and housing the upgraded facility. The agreement speaks of cooperation on a range of issues: research, technology, energy, environment, education, tourism, traffic planning and trade. The actual cooperation on these specific areas is to be coordinated through a 'joint committee',[63] and successful cooperation will thus essentially depend on the effectiveness of this body. In any case, the declaration arguably holds the promise of reducing Greenlandic dependence on Denmark and potentially facilitating the ultimate goal of self-determination.[64] The agreement also played an important role regarding the domestic constituency of the Greenlandic politicians. With the agreement in hand, the Greenlandic politicians were free to argue that they got something out of their new right to negotiate bilaterally with the US.

The combination of the explicit threat of leaving the island and the possibilities for Greenland associated with the agreement on economic and technical cooperation in the end paved the way for a compromise. However, the difficulties in reaching that agreement can in fact be seen in its rather vague text. Although the agreement mentions a range of issues suited for cooperation, how this cooperation is to function in reality and whether it will actually confer any technical or economic benefits to Greenland is yet to be seen.[65]

In this light, the outcome of the tough and long negotiations could be seen as a good result for the US. By initially leaving the initiative to the other parties and then applying mounting pressure, the US in the end obtained its permission to upgrade the radar at Thule, without paying too much in return. Seen from the perspective of the US, conveying an independent role in the future negotiations concerning Thule Air Base upon the semi-autonomous mini-state of Greenland and giving the Greenlanders what amounts to a liaison officer on the base, in return for the upgrade, seems like a good deal.

However, the political cost might rise in the future. The very intangibility of the agreement on economic and technological cooperation that made it possible in the first place, might actually be of political value to the Greenlanders in the time to come. The fact that the agreement can be interpreted in various ways entails that it will be easy for the Greenlanders to accuse the US of non-compliance – of not living up to its promises. Thus accordingly, the US might very well in the future find itself in a situation parallel to that of the Danish government. By accusing the US of not living up to Greenlandic definitions of American promises, the Greenlanders can perhaps use the agreement on economic and technological cooperation to pressure the US for further benefits. The politics of embarrassment could become a viable strategy towards the US as well. Thus the cost of the upgrade for the US might indeed show itself further down the road, and it will be difficult for the US not to deliver on the agreement. However, at least for the time being, no problems have arisen, and both parties can be satisfied.

The same holds for the Danish government. As stated earlier, reaching an agreement was the main objective of the Danish government – partially as a consequence of the pressure put on it by the American strategy of non-involvement. But the agreements also make life easier in the Danish administration. The domestic Danish–Greenlandic issue of Thule is effectively closed. The Greenlanders got the influence they wanted, and by conferring sovereignty on the Greenlanders any future Greenlandic concerns about the base at Thule are put in the hands of the Greenlanders themselves. Denmark can now – as was seen in the second round of negotiations – legitimately pursue the strategy of non-involvement in relation to Thule.

The announcement of the final agreements paving the way for the missile defence upgrade effectively marks the end of the process, and we can now turn to the insights provided by the case. The following will, before we turn to the actual conclusions, summarize the process from the point of view of all three parties, and then briefly discuss the end result in total.

The end result

The symbolic finalization of the upgrade negotiations took place on 7 August 2004, when Colin Powell – as the first US Secretary of State – visited Greenland where he, together with his Danish counterpart and the Greenlandic Deputy Premier of the Home Rule Government, signed the above-mentioned documents. The atmosphere at the signing was cheerful, and all parties seemed happy to complete the process. And, as the preceding pages have shown, all parties could indeed be satisfied in that they all obtained an acceptable result. In short, the US obviously received its permission to upgrade the radar, thus also confirming its close relations to Denmark (and Greenland). Greenland was formally accepted as an actor, thus increasing its influence as well as its independence. Denmark negotiated an agreement in a difficult setting, effectively balancing domestic constraints with outside pressure, the result of which might be to close the domestic issue of Thule.

However, even though in the end a result was reached, the previous analysis has shown that it was by no means easy. The process was complicated by the existence of the Greenlanders. Without the Greenlanders, and the distinct history of Thule Air Base, a decision to allow the US upgrade would have been made by Denmark at least as quickly as was the case with the UK. The history of Thule makes the issue of missile defence an internally contentious one in the Danish Realm. Greenlandic involvement is necessitated by this history, and it is accordingly this history that structures the process. It limits the choices of the other actors, especially Denmark, but in turn also the US, and the history thereby greatly contributes to the establishment of the strategies of all three actors. Both the US and Denmark have to take the Greenlanders into account. The distinctiveness of the US strategy in the negotiations can thus to a large part be ascribed to the particularities of the domestic situation in the Danish Realm.

Actually, we can see how this history structures the negotiations and thus to a large extent structures the strategies of the actors. First, history in itself is what makes the Greenlanders an actor. Further, it is obviously in order to deal with this issue that the negotiations are split into two bilateral negotiations. The Greenlanders become important in both negotiations, and it is their striving for influence and independence that structures how the negotiations proceed, and what issues are on the agenda, and indeed what issues are not on the agenda. What is particular in this process is in fact the very minor role that missile defence plays as such. What is at stake is not missile defence, but the trade-offs that missile defence entails on other areas. The discussions concern themselves with missile defence versus non-missile defence issues.

Both Denmark and Greenland agree early on what should be on the agenda. The Greenlanders state their case; missile defence is an opportunity to achieve further formal influence and independence vis-à-vis in particular Denmark. On the other hand, Denmark is not unwilling to compromise on the issue. Thus early on in the process the Greenlanders focus their effort on formal independence from Denmark instead of using their moral authority to directly influence any aspects of missile defence in general as well as on Thule. The Greenlanders choose, as their primary demand, future formal sovereignty on security policy instead of concrete, real and immediate influence on the missile defence project as such.[66] This effectively means that the subject of missile defence becomes a sideshow in the negotiations. There are good reasons to make this trade-off, however. First, Danish authority on Greenland is seen as an important and contentious issue on the island, and increasing Greenlandic influence and independence is considered a key political project. Second, as we have seen, the Greenlanders have available the powerful strategy of embarrassment for achieving this goal. This trade-off, however, still has the effect of closing off the issue of missile defence as such. Instead of negotiating terms and demands in relation to missile defence proper, other issues were at stake. Instead of saying 'yes, but' in relation to missile defence, the negotiations were about a 'yes, but' in relation to what Denmark, and subsequently the US, were willing to concede to the Greenlanders on other issues.

The result of this for the US is that the negotiations take place on parallel issues. As in the example of the Danish government White Paper, what was important was not missile defence as such, but other issues. Even though commentators critically question the reasons and intentions of missile defence, the consequence of missile defence deployment is not the main issue. The main issue is what trade-offs can be achieved. This undoubtedly works for the benefit of the US. Instead of being pressured into a long and potentially uncomfortable discussion on the merits of missile defence as such, and having to deal with a demand for concessions on this very central issue, the US can instead give in on issues unrelated to missile defence. The benefits for the US entailed in this particular dynamic of the negotiations can be visualized by comparing the Canadian debate. In Canada, cooperation on missile defence has at times been conditioned on the promise that the project would not lead to the militarization of space[67] – an issue that the other chapters of the present volume indicate would be difficult for the US to reach a compromise on. Accordingly, Canadian commitment to missile defence has not been especially forthcoming, and in February 2005 Canada announced that it would not be joining the system. The dynamics that made the internal issues in Denmark central, in turn made the process easier for the US; it took missile defence as such off the agenda, and in essence provided the US with an obvious opportunity for making a trade-off on a totally parallel issue. This, on the other hand, does not imply that the process was easy; it took time and a range of manoeuvres in the negotiations to reach an agreement.

In the end, however, even though the process was long and hard, and the negotiations dragged on, straining the relationship of the three parties,[68] the US got its upgrade. Thus the US presence on Greenland is maintained and work on the base integrating it into the missile defence is scheduled to be completed by 2006.

Conclusions

Making generalizations from a single case is hazardous, especially in this particular case, as the process leading to the upgrade of Thule Air Base is in many ways unique. Nevertheless, elements of the case are likely to be found elsewhere, and the case thus provides insights into a number of issues and problems associated with any future bi- or multilateral missile defence negotiations the US might find itself in. The case further provides insights into some issues of more general interest.

First, although the history accounted for in the above is largely specific and distinct for the base at Greenland, it nonetheless shows how domestic issues can constrain governments and greatly affect foreign policy. In this particular incident, the interesting aspect is the 'power of the weak'. By skilfully using the history of Thule, materially speaking, weak Greenland succeeds in achieving a position of moral authority in the debates and negotiations with Denmark concerning the actual upgrade. The power following from this historically conditioned moral authority in turn means that the Greenlanders make their presence felt in the missile defence negotiations. Overall, the result of the negotiations is that the relationships between Denmark, Greenland and the United States change. The Home Rule Government emerges as a new almost state-like actor. The political arrangements surrounding Thule Air Base are thus complicated by the arrival of this new actor. These issues – the importance of history and morality as well as domestic pressure – show how international negotiations can be a complicated business.

Other insights can be drawn from this case with more general implications for the US missile defence plans. The case shows how the US has to choose various strategies in various cases. What works in relation to this case might face the risk of failing with other actors. The case in fact showed that the US chooses a specific strategy in the case of Greenland. The strategy of non-involvement had both benefits and drawbacks in this particular case. First it was responsible for conveying a positive impression of the US in the first round of negotiations. The US presented itself almost with modesty, attentive to the domestic demands of Denmark and the wishes of the 'weak' Greenlandic actor. This was probably a wise strategy as a more assertive and pushy US would undoubtedly have generated counter-pressure in the form of left-wing opposition in Denmark as well as being accused of acting undemocratically and immorally by the Greenlanders. Later in the process the drawbacks of this strategy showed themselves, as the negotiations went into deadlock. However, maybe it was exactly the previous attention to the wishes of the other actors that in the end made the rather hard – but out in the open – pressure applied by the US possible. Overall the strategy worked in the end. Specific settings require specific strategies; that is the lesson of this case.

However, nothing guarantees that the US will not in the future run the risk of again being drawn into long and complicated processes with potentially disrupting effects. For instance, opposition was voiced concerning the Fylingdales upgrade as well,[69] but the power of the opposition arguments was – as a matter of a different history – not as strong as that of the Inuit of Greenland. The prospect of interference from various local oppositions might indeed also be the case if the US administration decides to deploy interceptors on, for example, European soil.[70] Both concerns about launch decisions and the potential risk of fallout are issues that could generate local concerns – even to the point of constraining their own government. Further, in areas with higher tension, following the logic of the security dilemma, some domestic groups might see the actual deployment of interceptors as having a direct negative impact on their own security.[71]

Where and whether these local logics will show themselves is hard to say – and that is exactly the point. As more actors become involved in various ways, the more unpredictable are the reactions. The future challenge of developing and deploying a global missile defence system incorporating friends and allies can only be met successfully with increased attention to distinct and varying local issues by the American administration. What is certain is that the proliferation of its missile defence plans will inevitably draw the US into a range of localized negotiations in the future, and although these negotiations require skill and political attention to local issues, they also hold potential benefits for the US. The result of this particular case, at least, has not been a deterioration of the relationship of the three parties, on the contrary. In addition to obtaining permission to upgrade the radar at Thule, the US has drawn the other parties closer to itself.

Notes

1 Israel, Germany, Japan and Italy, among others, have for some time worked together with the US on missile defence. More recently Australia has announced its participation as well, clearly in line with President Bush's aspiration to involve 'friends and allies'; see for example George W. Bush, 'President Announces Progress in Missile Defence Capabilities', 17 December 2002. Online. Available HTTP: ⟨http://www.whitehouse.gov/news/releases/2002/12/20021217.html⟩.

2 George W. Bush, 'President Delivers State of the Union Address', 29 January 2002. Online. Available HTTP: ⟨http://www.whitehouse.gov/news/releases/2002/01/20020129-11.html⟩.

3 On the Arctic strategy and the historical importance of Thule Air Base, see DUPI (Dansk Udenrigspolitisk Institut), *Grønland under den kolde krig: Dansk og amerikansk sikkerhedspolitik 1945–68*, Copenhagen: DUPI, 1997, pp. 114–117, 219–239; Clive Archer, 'The United States Defence Areas in Greenland', *Cooperation and Conflict*, 23, 1988, pp. 123–134.

4 Statement by Lucas Fischer, Deputy Assistant Secretary of State for Strategic Affairs, during 'Conference on Missile Defence and its Implications for the Global Order', 25 April 2001, at the Danish Parliament.

5 See for example Teemu Palosaari and Frank Möller, 'Security and Marginality: Arctic Europe after the Double Enlargement', *Cooperation and Conflict*, 3, 2004, p. 273.

6 Clive Archer, 'Greenland, US Bases and Missile Defence: New Two-Level Negotiations?', *Cooperation and Conflict*, 2, 2003, p. 133; DUPI, op. cit., pp. 273–275; Poul Villaume, *Allieret med forbehold. Danmark, NATO og den kolde krig. Et studie i dansk sikkerhedspolitik 1949–1961*, Copenhagen: Eirene, 1995, p. 390; Nikolaj Petersen, 'Negotiating the 1951 Greenland Defence Agreement: Theoretical and Empirical Aspects', *Scandinavian Political Studies*, 21, 1998, p. 2.

7 The most cited example is the continuous lower than NATO average Danish defence spending. See for example Archer, op. cit., p. 133.

8 Most notably the decision to break away from the EU (EEC at that time) in 1985. See further Permanent Secretary of State for Foreign Affairs Friis Arne Petersen, 'Rigsfællesskabet og det internationale arbejde for oprindelige folk', *Udenrigs*, 2, 2001, pp. 75–83.

9 Not coincidentally just before the passing of the constitutional amendment conferring Danish citizen rights on the Greenlandic population.

10 Especially through a book published in 1987: Jens Brøsted and Mads Fægteborg, *Thule – fangerfolk og militæranlæg*, Copenhagen: Akademisk Forlag, 1987, pp. 50–64.

11 A process that was only settled in 2003 by the Danish Supreme Court, which acknowledged the unlawfulness of the government action. The Inuit plaintiffs were satisfied neither with the assumptions behind the Court's ruling nor the size of the compensation, however. Therefore the legal aftermath is not yet at an end, as the Inuit of Thule have appealed the case to the European Court of Human Rights.

12 See for example P. Clæsson (ed.), *Grønland, Middelhavets Perle. Et indblik i amerikansk Atomkrigsforberedelse*, Copenhagen: Eirene, 1983; Archer, op. cit., pp. 139–142.

13 This is extensively covered in DUPI, op. cit., especially pp. 277–302.

14 Hans Mourizen, 'Thule and Theory: Democracy vs. Elitism in Danish Foreign Policy', in Bertel Heurlin and Hans Mouritzen (eds), *Danish Foreign Policy Yearbook 1998*, Copenhagen: DUPI, 1998, p. 81.

15 Press Statement, US Department of State, 'Secretary Powell Meets with the Foreign Minister of Denmark and the Minister for Economic Affaris and Vice Premier of the Greenland Home Rule Government', 18 December 2002. Online. Available HTTP: ⟨http://state.gov/r/pa/prs/ps/2002/16104.htm⟩.

16 Secretary of State for Defence Geoff Hoon, 'MOD Responds to US Request to Upgrade RAF Fylingdales', Press Notice no. 025/03, 5 February 2003. Online. Available HTTP: ⟨http://news.mod.uk/news/press/news_press_notice.asp? newsItem_id=2271⟩.

17 Press notice, Danish Ministry of Foreign Affairs. Online. Available HTTP: ⟨http://www.um.dk/da/menu/Udenrigspolitik/FredSikkerhedOgInternational Retsorden/Missilforsvar/Thule-radarensRolle/ AftalekompleksitilknytningtilopgraderingafThuleradaren.htm⟩.

18 See for example *Berlingske Tidende*, 17 July 2001, section 1, p. 10; *Politiken*, 6 September 2001, section 1, p. 1.

19 Press Statement, US Department of State, op. cit.; Deputy Assistant Secretary for Strategic Affairs, State Department, Robert Lucas Fischer, Statement during public hearing conducted by the Foreign Policy Committee of the Danish Parliament, 25 April 2001.

20 On the positive attitude of the Danish government to missile defence before the request, see *Politiken*, 6 November 2001, section 1, p. 1; on the political promise to integrate the Home Rule Government in the decision, see *Berlingske Tidende*, 15 October 2001, section 1, p. 8; on the general will to extend the influence of the Home Rule Government on security policy, see *Politiken*, 2 October 2002, section 1, p. 4.

21 *Berlingske Tidende*, 18 December 2002, section 1, p. 8.

22 Robert Lucas Fischer, op. cit.

23 Danish Foreign Minister Per Stig Møller, quoted in the web edition of *Politiken*, 17 December 2002. Online. Available HTTP: ⟨http://politiken.dk/ VisArtikel.sasp?PageID=247938⟩.

24 Ronald Niezen defines this strategy and the concept in the following terms: '[T]he use of the media, political lobbying and public relations campaigns to highlight the abuses of the state – has been effectively applied by... indigenous organizations in order to encourage government recognition of indigenous peoples' distinct claims.' See 'Recognizing Indigenism: Canadian Unity and the International Movement of Indigenous Peoples', *Comparative Studies in Society and History*, 1, 2000, p. 143.

25 Greenland enjoys substantial financial support as well as wide-ranging political autonomy from the Danish state. But, as stated earlier, this is not the case in security and defence policy. These policy areas have been conducted exclusively from Copenhagen. For further aspects in general about a distinct Danish liberal moralistic foreign policy profile, see for example Christine Ingebritsen, 'Norm Entrepreneurs: Scandinavia's Role in World Politics', *Cooperation and Conflict*, 1, 2002, pp. 11–23; Lene Hansen, 'Sustaining Sovereignty: The Danish Approach to Europe', in Lene Hansen and and Ole Wæver (eds), *European Integration and National Identity: The Challenge of the Nordic States*, London: Routledge, 2002; Peter Lawler, 'Scandinavian Exceptionalism and European Union', *Journal of Common Market Studies*, 4, 1997, pp. 565–594.

26 On the strength of moral arguments and moral authority in international relations in general, see Rodney Bruce Hall, 'Moral Authority as a Power Resource', *International Organization*, 4, 1997, especially p. 594.

27 Report from the Danish government, *Missilforsvar og Thule-radaren, Redegørelse fra regeringen*, Copenhagen: Ministry of Foreign Affairs, March 2003. All following quotes by Danish and Greenlandic actors are translated from Danish by the author.

28 Ibid., p. 8.

29 Ibid., p. 30.

30 Ibid., pp. 26–28.

31 Ibid., p. 33.

32 *Politiken*, 24 April 2003, section 1, p. 7.

33 Ole Wæver in *Information*, 14 May 2003, section 1, p. 8.

34 Danish Prime Minister Anders Fogh Rasmussen, quoted among others in *Politiken*, 23 January 2003, section 1, p. 2.

35 See for example David Wright from the Union of Concerned Scientists quoted in *Information*, 11 March 2003, section 1, p. 4; and Sir Timothy Garden, 'if a missile defence system is developed, then it is obvious that these places [Fylingdales and Thule] will be important targets for an aggressor against the USA. Normally, the first thing you do is to remove the eyes of the system', statement during public hearing conducted by the Foreign Policy Committee of the Danish Parliament, 25 April 2001.

36 Lars Emil Johansen, quoted in *Politiken*, 5 March 2003, section 1, p. 5.

37 In addition to two open hearings and a number of debates in the Danish Parliament, the Foreign Minister is quoted extensively on the issue; see for instance *Information*, 25 April 2003, section 2, p. 15; *Politiken*, 5 March 2003, section 1, p. 5.

38 See for instance *Politiken*, 10 March 2003, section 1, p. 4 on the opposition view; editorial in *Jyllands-Posten*, 18 December, 2002, section 1, p. 8 on the legitimacy of the historically based Greenlandic demands.

39 Joint Statement by the Foreign and Security Committee of the Home Rule Authority, issued 15 November 2002. Online. Available HTTP: ⟨http://dk.nanoq.gl/udskriv.asp?page=nyhed&objno=41050⟩.

40 According to some Greenlandic politicians Denmark has already received 'payment' for making Thule Air Base available for the American missile defence plans. This ranges from promises of subcontracts to Danish companies to including southern Denmark in the coverage of the defence free of charge. See Lars Emil Johansen in *Jyllands-Posten*, 23 November 2002, section 1, p. 9 and statement by Greenlandic member of the Danish Parliament Kuupik Kleist during a debate in the Danish Parliament, 29 April 2003.

41 The only reference to the inhabitants of the island in the original agreement was that any contact between the local population and US personnel in the base areas should be avoided 'with every effort'. That was changed only in 1986 with an amendment permitting the population work opportunities, etc., on the bases. The full title of the agreement is 'Agreement between the Government of the United States of America and the Government of the Kingdom of Denmark, pursuant to the North Atlantic Treaty, concerning the defence of Greenland'. Online. Available HTTP: ⟨http://www.yale.edu/lawweb/avalon/diplomacy/denmark/den001.htm⟩. On this issue see Article VI of the agreement.

42 Mostly because of American opposition, apparently concerned with the prospect of getting a renegotiated agreement through Congress. The Danish government thus argued that this would result in a long and unpredictable process, over what would probably be seen in Congress as a technicality. Furthermore, the process would probably result in loss of both Danish and Greenlandic goodwill in the American administration, thus damaging the close relationship without any prospect of success. Apart from that, the United States has in principle expressed understanding of the need to discuss the principles regulating the American presence in Greenland, and has appeared willing to discuss these issues with the Greenlandic Home Rule Authorities. See Assistant Secretary of Defense J.D. Crouch II, quoted in *Politiken*, 7 May 2003, section 1, p. 4.

43 Joint Declaration in Principle between the Danish Government and the Greenlandic Home Rule Government on the Involvement of Greenland in Foreign Policy and Security Policy. Online. Available HTTP: ⟨http://dk.nanok.gl/nyhed.asp?page=nyhed&objno=54781⟩.

44 Announcement by the Danish Minister of Foreign Affairs and the Premier of the Greenland Home Rule Government. Online. Available HTTP: ⟨http://dk.nanok.gl/nyhed.asp?page=nyhed&objno=54781⟩.

45 Member of the Greenlandic Home Rule Parliament Augusta Salling, quoted in *Jyllands-Posten*, 15 November 2002, section 1, p. 5.

46 On the American position on this issue, see for example Assistant Secretary of Defense J. D. Crouch II: 'We seriously appreciate the Danish and Greenlandic wishes for a renegotiation of the 1951 Agreement, and we will find ways to do it', quoted in *Politiken*, 24 April 2003, section 1, p. 7.

47 The following discussion of the US strategy towards the UK and Canada draws in large part on the analysis presented in Andrew Richter, 'A Question of Defence: How American Allies are Responding to the US Missile Defence Program', *Comparative Strategy*, 2, 2004, pp. 143–172.

48 Ibid., pp. 149, 154–155.

49 Ibid., p. 151; on the aspect of defence industry cooperation see further *Washington Post*, 19 October 2003, Section A, A27.

50 See for example *National Post*, 23 August 2004, section A, p. 4 on the costs for the Canadian defence industry if the government decides not to participate. Further on the British case see Richter, p. 150.

51 Ibid., op. cit., p. 155.

52 See especially the statements by Assistant Secretary of Defense J.D. Crouch II at the public hearing in the Danish Parliament on 23 April 2003, quoted in, among others, *Politiken*, 24 April 2003, section 1, p. 7; Assistant Secretary of Defense Mira R. Richardel, quoted in *Berlingske Tidende*, 26 March 2004, section 1, p. 15.

53 See for instance *Berlingske Tidende*, 30 April 2004, section 1, p. 11.

54 Greenlandic member of the Danish Parliament Lars Emil Johansen in *Weekendavisen*, 6 February 2004.

55 Support for conceding to the US request was not forthcoming in the Greenlandic population, hence the importance of showing that giving an affirmative answer would at least pay something in return. See opinion polls quoted in *Information*, 25 April 2003, as well as in *Weekendavisen*, 2 May 2003, showing approximately 42 per cent of the Greenlanders opposing the upgrade with 23 per cent undecided and only 35 per cent in favour.

56 *Politiken*, 7 February 2004, section 1, p. 5; ibid., 21 February 2004, section 1, p. 2; *Berlingske Tidende*, 26 March 2004, section 1, p. 15; ibid., 30 April 2004, section 1, p. 11, all account and discuss mounting US frustration with the lack of progress in the negotiations.

57 Apparently an agreement had to be reached before the passing of the US budget for financial year 2005. *Berlingske Tidende*, 26 March 2004, section 1, p. 15.

58 The possibility of pursuing other options, whether alternative locations or alternative technology, was being voiced as a potential option, for example by Assistant Secretary of Defence J. D. Crouch II, quoted in *Fyens Stiftidende*, 27 April 2003. That would be bad for the Greenlandic economy. The American presence on Greenland provides jobs – a scarcity in the Arctic – and contracts to Greenlandic firms provide revenue to the Home Rule Government. On the economic benefits of the base for Greenland, and consequently also the consequences if the US was to leave Greenland, see Preben Bonnén in *Jyllands-Posten*, 9 January 2003, section 1, p. 3.

59 See Article IV of the agreement. Online. Available HTTP: ⟨http://www.um.dk/NR/rdonlyres/EF0F61CB-500B-491A-9465-C9E95B6BA0B4/0/ModerniseringafForsvarsaftalenaf1951.DOC⟩.

60 See the preamble to the agreement.

61 Article III, paragraph 1, section 3.

62 The agreement can be accessed online. Available HTTP: ⟨http://www.um.dk/NR/rdonlyres/45B44E9B-8BF9-4D66-9112-9E314CCAB9E2/0/Oekonomiskte kniskerklaering.doc⟩.

63 Paragraph 1, section 1.

64 As an example of the issues under consideration, it can be mentioned that Greenland and the US together with Denmark have initiated new cooperation in Arctic research; see ⟨http://dk.nanoq.gl/nyhed.asp?page=nyhed&objno=67576⟩.

65 Even though the agreement mentions all these various issues for cooperation, the actual relationship is dependent on, for instance, 'the availability of funds in accordance with national laws and procedures' (see Paragraph 3 of the agreement); moreover, much depends on the yet to be developed cooperation in the 'joint committee'. Thus, effective cooperation depends on the political will of all participants.

66 The opinion that Denmark and especially Greenland should have tried, if only in a small way, to influence the US missile defence plans by making a positive reply dependent on, for instance, a certain future architecture, etc., has been voiced in the media. In effect the Greenlanders, instead of saying yes, and getting concessions in relation to other issues, should say 'yes, but', implying that they should formulate some demands narrowly tied to missile defence. See for instance op-ed. by Ole Wæver in *Information*, 14 May 2003, p. 8.

67 Richter, op. cit., p. 156.

68 See for instance *Berlingske Tidende*, 30 April 2004, section 1, p. 11.

69 For an overview, see Mark Bromley and David Grahame, 'NMD: Overview of the Political Debate in the United Kingdom', Basic Notes, 1 December

2001. Online. Available HTTP: ⟨http://www.basicint.org/pubs/Notes/2001NMD_UKdebate.htm⟩.

70 On the prospect of missile defence launch sites in Europe, see for example *Christian Science Monitor*, 13 July 2004.

71 Obvious candidates are South Korea or Taiwan; for the case of Taiwan, see Nicole C. Evans, 'Missile Defence: Winning Minds, Not Hearts', *Arms Control Today*, 5, 2004, pp. 48–55.

CONCLUSION

Bertel Heurlin, Sten Rynning and Kristian Søby Kristensen

In an era of new threats and global change, a political leader runs great risks: he may stake innovation on technologies that turn out not to work; he may run up against an enemy who more perceptively exploits the energies of change; and he may lose his friends who find greater potential in their own political arrangements. Missile defence is one component in a US strategy designed to ensure that none of these risks are realized.

As with any design, its real test comes not from its inner coherence – which is likely to be impeccable – but from its ability to deal with real challenges. This book set out to contribute to our understanding of the way in which the policy of missile defence might cope with regional dynamics, and thus whether missile defence contributes to the revitalization of America's imperial fabric.

The chapters of the book are marked by their attentiveness to local, regional circumstances, and it may be difficult to generalize from this multitude of actions and actors. Still, in light of the risks just mentioned, it is possible to offer a general assessment of the US design and notably the likelihood that it will win the US more friends or foes.

America's search for security and primacy

In the Introduction we posed the deceptively straightforward question of why. Why deploy missile defence? The question raises a host of new questions related to the ulterior motive of the project – the range of threats with which the US is faced and also its vision for continued global leadership.

In much of the existing academic literature as well as in the public debate, a central dividing line runs between those who believe the system is directed against rogue states and those who believe we are seeing only the beginning of a grander defence system that will perpetuate US military superiority – a renewed Star Wars project. The three chapters examining the history of strategy and US policy lead us to the conclusion that this dichotomy is misleading: missile defence is about countering new threats in order to perpetuate leadership. Granted, because of the technological challenges involved we do not know whether missile defence will actually work, but the evidence on political motivation seems clear.

Cronberg and Heurlin recognize that the threat from rogue states is important in the American missile debate, but they place the accent on deeper sources of motivation related to superpower status and military-technological innovation. Missile defence can in fact be compared to the logic of the Cold War nuclear arms race, as Heurlin points out: the purpose of both is to obtain political advantages by their non-use. The weapons will have succeeded if they prevent war. Thus, in the terminology of Agrell, missile defence is not only about defence but also deterrence and dissuasion. And we know that deterrence and dissuasion form part of a larger strategic policy, which thus supports the conclusion that missile defence is as much about US grand strategy for the future as defence today. Cronberg and Heurlin connect the present to the future by underscoring the research and development aspect of missile defence, in particular the role of new technology in outer space. Innovation today is the foundation of tomorrow's advantage, and innovation is therefore a central component in the effort to dissuade potential peer competitors from engaging in a new test of strength.

All this is not to deny the rationale of directing new defences against new threats. There is no doubt that the desire to protect both US territory and deployed troops from the dangers stemming from the proliferation of WMD and the emergence of new radical ideologies has played a major role in the domestic American debate. The origins of the debate can be traced to the early 1990s when the Soviet Union collapsed and when new regional dangers emerged: it took no stretch of the imagination to connect past stability and new uncertainties and thus question the assumption that state leaders can be rationally deterred. The 1991 Gulf War concretely saw US troops threatened by ballistic missiles potentially armed with WMD, and with the new unipolar position of the US it was clear that the situation would arise again in the near rather than distant future and also that new types of missile threats might develop.

Foreseeing new threats is a delicate balance between anticipating trends and mistaking imagination for reality. This delicate balance was perhaps fairly easy to walk for the Clinton administration during the years when failed states like the former Yugoslavia, Somalia and Haiti dominated the security agenda, but the balance provoked a contentious debate by the mid- to late 1990s when some political parties saw in the behaviour of illiberal regimes around the world the offsprings of political revision. These revisionists, the argument ran, had already drawn the lesson from the Gulf War that the US was best constrained by the threat of real WMD capabilities, and they had set out to acquire the relatively inexpensive missile technology to transport WMD. Candidate and later President George W. Bush advocated this position, and the terrorist attacks of September 2001 tipped the political balance in its favour. Still, as the analysis also points out, it is important to keep in mind not only the domestic politics of missile defence but also structural, international imperatives. In a bipolar setting, a superpower may come to adhere to a policy of stability through deterrence or mutual vulnerability; in a unipolar setting, the remaining superpower is not likely to value mutual vulnerability – after all, unipolarity implies that one is strong and thus able to reduce its vulnerability – but rather the degree of freedom of action it can acquire to establish and protect governing principles for international relations. Put differently, and in the words of the US national security strategy, deterrence under unipolarity enables 'nuclear blackmail'[1] whereby small actors tie down the global Gulliver.

Current new threats and future American leadership are thus the twin motives for missile defence. However, as this book has been organized to demonstrate, regional politics complicate the business of implementing the missile defence blueprint. Why, for instance, should the US offer its allies a share in the missile defence architecture, including a say in parts of the overall command and control system, if the US simultaneously considers missile defence one of several means to maintain international primacy? Moreover, why, in a prism of primacy, would the US focus so heavily on just one threat – rogue nations – when other and larger nations are more likely in the long run to be able to challenge US primacy? The brief answers are that the US needs allies to help set up the global missile defence infrastructure and also to assist in the campaign against regional revisionism, and moreover that the US does not want to provoke potential peer competitors by explicitly tying their potential power to the US defence architecture. These designs make sense given the interrelated nature of regional and international politics but, as the analysis has shown, the world does not always bend to US policy and thereby creates unforeseen consequences. We shall briefly consider this dynamic from the perspective of allies, rogue nations and peer competitors.

The allies

The European allies, Rynning concludes, are likely to strengthen their support for missile defence, but the support will nevertheless come attached to a demand for close allied consultations and a stronger American commitment to a wide security engagement where, in the eyes of European governments, military forces are more balanced by political dialogue and economic incentives. This conclusion should altogether be of comfort for American decision-makers.

Why is it that Europeans generally in the 1990s accepted the need for theatre missile defence but – sometimes adamantly – opposed strategic missile defence? The answer is likely that Europeans believe they can deter state leaders who gain the capacity to threaten European territories but that they have concluded, from the 1991 Gulf War and the many other interventions undertaken by various European coalitions through the 1990s, that their deployed troops are increasingly coming under the threat of missile attacks. Moreover, the distinction was an inherent part of the ABM treaty and thus the regime stabilizing Russian–American strategic relations. European governments, neighbours to Russia and its 'near abroad', were and continue to be sensitive to strategic changes that might increase Russia's sense of weakness and thereby provoke it to rely on the one military measure it continues to master – offensive nuclear weapons. The distinction between theatre and strategic defences thus made 'functional' sense to many Europeans: if strategic deterrence worked, why fix it? To this one might add that European states are regional powers at best and thus that the concern with future supremacy is less widespread in Europe.

This latter point should not be stretched too far because European governments, like the American, are concerned about the fate of the global institutions that underpin the status quo. They reacted firmly to the terrorist attacks in September 2001 and they underpin NATO's growing engagement in Afghanistan. Their concern with regional, rogue nations, as Rynning underscored, informs the convergence of European and American security agendas, including on the issue of missile defence. This rapprochement has no doubt been assisted by the docile Russian reaction to the US decision to withdraw from the ABM treaty and thus herald a new era of strategic policy. Missile defence is now one of the subject-matters for the NATO–Russia council formed in 2002, and although Russia remains sceptical about the strategic implications of American designs and is likely to implement a number of strategic countermeasures, as Kassianova argued, the continuing dialogue on the issue in NATO is of comfort to European allies.

The US is advised to carefully consider the local circumstances of allies, is the lesson from the chapters of Dörfer, Hansen, Dyvad and Kristensen, because they are able to support US policy only to the extent that their national situation permits it. Dörfer generally categorizes Scandinavia as a region where the issue of new threats and the need for new countermeasures is suppressed – one of the categories in Agrell's overview of strategic concepts. Scandinavians, it would seem, are too little exposed to the developments engulfing the US to support the missile defence concept. Still, important intra-regional differences should be underscored. It would seem that engrained traditions of neutrality and also proximity to a great neighbour like Russia motivates the scepticism found in Sweden and Finland. Norway and Denmark are in contrast warming to the issue: both are founding members of NATO, a critical element to keep in mind according to Dörfer, and, as Kristensen underscores, Denmark is hosting one of the key radar installations and, in continuation of its Cold War practice, sought to blend principled support and pragmatic gain in the negotiations that resulted in an agreement to allow for an upgrade of the radar site.

Allies such as Taiwan, Japan and Israel have enthusiastically embraced missile defence because, in line with the regionalization perspective of the chapters, they are located geographically in regions with no lack of missiles, WMD capabilities and dubious political motives. This does not imply, however, that the US can furnish missile defence to these allies and thus obtain stability. As Dyvad points out, the attempt to insulate allies from outside aggression can be perceived as a hostile act by regional great powers such as China, and the US must therefore balance the local interests of its allies with its overall national wish for strategic predictability and stability. The allies are therefore likely to obtain local missile defence capability, if at all, with strings attached. This was a central theme in Birthe Hansen's assessment of the Middle East and the fact that Israel is the only regional nuclear power. If missile defence is to contribute to stability, then a greater package of stability measures must be put on the table, she argued, continuing that Israel must accede to regional arms control arrangements.

The rogues

The states bent on political revision for reasons of political, religious or economic gain are key to the missile debate and also to assessments of whether missile defence actually is able to influence threatening behaviour.

A critical aspect to keep in mind is the extent to which these states are motivated by regional questions and not merely the desire to challenge the global hegemony of the US. Indeed, it would be irrational to directly challenge the US, and regional motives no doubt play a prominent role in the two regions containing most rogue nations, so labelled by the US administration, and other nations of concern: Pakistan is not directing its missiles against the US but against India, North Korea against South Korea, Iran against Iraq or Israel, and so on.

The US is drawn into these dynamics *qua* its global leadership role: its interests may conflict with those of regional powers, leading to the American brandishing of a particular power as a rogue nation threatening missile attacks on the US. This conflict of interests is built into the current international structure of unipolarity, one might note, and rather than searching – in vain – for a way to abolish these conflicts, it would be more fruitful to search for ways of stabilizing them. It may be, naturally, that a particular conflict can be transformed and thus solved – for instance, if North Korea underwent a regime change and agreed to the unification of the peninsula – but conflicts between the unipole and regional revisionists will continue.

Thus, irrespective of whether revisionists are mostly motivated by regional questions, they will enter into conflict with the global power, the US, as revision becomes more likely and runs against the grain of American thinking. Revisionists or rogues will generally lack the means to match US nuclear capabilities, forcing them to consider one of two options: acquiescence or asymmetric response. With the first option the rogue would concede victory to the US, acknowledge that missile defence removes the rationale for its sparse missile capability, and seek a transparent disarmament agreement: Libya has done so. However, acquiescence does not need to imply strategic defeat, it should be noted, because a concession to the US, however significant, could be made in order to further other regional ambitions linked to the country's influence among its neighbours. Acquiescence on missiles could for instance lead to renewed investments in conventional capabilities or, alternatively, a strategy of economic growth and dominance. The other option is to continue the attempt to deter the US, perhaps in the belief that a strategy of regional change can work only if the US global hegemony can be intimidated. Asymmetry could notably result in the threat of delivering WMD with other means than missiles, thereby rendering the missile defence impotent. Such means of delivery are numerous, ranging from the proverbial suitcase to containers transported to US cities via regular trading routes. The US, along with its allies, knows this and is developing defensive programmes in addition to missile defence intended to intercept such efforts – such as the Proliferation Security Initiative. What the option tells us, however, is that the game of tit-for-tat between the US and local revisionists is likely to be ongoing and involve a range of measures: missile defence may be a critical measure, but it is just one of many and its effects will depend on the greater political context.

The most pressing question in this context may be the case of Iran. As Birthe Hansen noted, it is probably too late to reverse the Iranian nuclear programme, and Iran will likely emerge as a nuclear power in the near to medium future. In Birthe Hansen's Waltzian assessment, Iran's accession to the nuclear club is not in and of itself destabilizing because Iran will be interested in its survival, and it will articulate an asymmetric doctrine principally to deter the US from interfering in Iranian affairs. Deterrence can thus be coupled with wider regional processes of change, Hansen argues. North Korea represents another case of rogue behaviour and an often-cited motive for the current limited missile defence capability, as Peter Dyvad demonstrated. North Korea quite clearly exploits its emerging nuclear capability asymmetrically, and everyone understands that the regime does not have the economic or industrial foundation to augment this capability beyond a bare minimum. Still, North Korea is determined to exploit the full political potential of its capability, and it is at this point that the question of local revisionists merges with the question of strategic rivals to the US.

Potential peer competitors

China is a potential peer competitor, and China may not want its relationship to the US to be upset by its small rogue neighbour, North Korea, as Peter Dyvad showed. North Korea's capability could harden the US determination to press ahead with the missile defence build-up, and it would in time incite China's neighbours such as South Korea, Japan and Taiwan to seek counter-capabilities – to go nuclear. China and the US could thus see eye to eye on the North Korean issue, but this is where the overall strategic context becomes critical. The US officially designates rogue nations as the *raison d'être* of missile defence and even agrees to share certain technologies with Russia, but still, the fine print of official documents reveals also the purpose of dissuading peer competitors.[2] China is certainly a recurrent country in the long-term projections emanating from the US Defense Department, and it is no coincidence that the Pentagon often refers to China and not Russia: the potential of its economic and social base to support new foreign policies is markedly stronger. This might also explain the ease with which the US has decided to collaborate with Russia on some missile defence issues, as Kassianova shows. It may also be the reason why it is difficult at present to categorize Russia or China as either rivals or collaborators on the issue of strategic doctrine.[3]

Russian leaders tend to see eye to eye with US leaders on the issue of international terrorism but are wary on the issue of missile defence. They are likely cooperating with the US in order to probe the American plans and potential and also to reap the diplomatic benefits of collaboration; but the essence of Russian policy is nevertheless the investment in countermeasures whose purpose is to ensure that the American shield will hold promise only against minor states and, by implication, not against Russia. This latter type of investment seems also to be the choice of China, whose strategic nuclear forces are moderate but growing.

China in fact represents a powerful illustration of the interrelatedness of regional and international questions. China's concern for the strategic implications of missile defence is probably moderate: the defence is not yet a proven capacity and China's international status is evolving. There is thus ample time to articulate a position on the strategic dimension, and China and the US have engaged in a type of diplomatic dialogue on the topic. China and the US thus worked together in the 1990s to make the Non-Proliferation Treaty permanent and to agree on terms for the Comprehensive Test Ban Treaty. Policy-makers may take comfort in the common US–Chinese interest in non-proliferation but they should nevertheless worry about the implications of regional questions because third actors maintain the capacity to upset Chinese–American relations. The ongoing six-nation negotiations on North Korea are merely a case in point: neither the might of the US nor the ties to China seem to hold decisive sway in Pyongyang. Taiwan is another joker. Taiwan wants missile defence and can appeal to the good conscience of the US in its effort to obtain it, but it may also manipulate the level of tension with China to put pressure on the US. The US must, as noted above, protect its allies but is also worried about the regional implications of introducing defence where deterrence previously worked. For instance, if Taiwan acquires a defensive capability that China perceives to be effective, then the Chinese leadership – whose primary goal will likely remain Chinese unification – could choose to launch a strategic challenge against the sponsor of Taiwan's defence, the US, and thus would unleash the type of strategic arms race that missile defence critics fear and US policy-makers strive to avoid.

On the crossroads of American security policy and regionalization

The restoration of the imperial fabric in thirteenth-century Europe fell to a string of popes and Holy Roman Emperors. Pope Boniface VIII was perhaps the most ardent advocate of the subjugation of earthly powers to his church, and his 'One Holy' doctrine inspired him to attempt to overthrow Philip IV of France in 1302. This dash for empire failed miserably, however: Boniface VIII was kidnapped and tortured by Philip IV and died shortly thereafter.

There is no equivalent to Philip IV in the current international domain – no earthly power able to unseat the leading power of the current international system, the US. This much is clear from the preceding analysis. But whether the current missile defence policy will reinforce a modern version of the 'One Holy' conception of global order, and whether the US in its reach for it will provoke its own downfall, is less clear. The analysis has highlighted the sources of US policy, the vision and ambition, as well as the distinct dangers inherent in various regions of the globe.

The first three chapters made the case why missile defence is not only about a particular new threat but rather a type of 'One Holy' world order. The US is tying the concept of defence to other strategic concepts, thereby invigorating efforts to maintain a competitive advantage in military affairs writ large, with the technological dimension of the revolution in military affairs and outer space being key dimensions. It makes sense for US policy-makers to design such a course of action: the US has a track record of benevolent leadership, albeit not an untainted one, and it stands to gain tremendous influence if its current position can be maintained.

The real question is whether it will work. The remainder of the book provided a first assessment. There is no present-day Philip IV because the potential peer competitors lack the material foundations for competition, because the US works fairly well with them on the issue of international terrorism, and because traditional US allies continue to support the US and not the peer competitors. The threat from rogue states is the most current threat today, which is reflected in the converging European and US security agendas. Rogue nations lack the power to usurp the 'One Holy' of the twenty-first century, but it is possible that they will gain in power or set off conflicts that will bring about a much greater challenge to American leadership. This likelihood raises some questions for further research.

The US and its allies must join forces to address the threat from rogue nations because the threat cannot be reduced to missile launches: it is a broad threat related to a crisis of legitimacy and capacity in the governing system of a particular country or region. It seems futile to debate whether the threat should be countered by military or other means: a combination is required. This opens an intriguing research question: in what ways do rogue nations differ, and in what ways are they thus susceptible to various countermeasures? It may be that the US in some cases is right that some rogue nations must be countered with firm military responses; it may also be that Europeans are right that some rogue nations must be dealt with in a broader process of political and economic change. But which comes first – security or development? And how can defensive military measures reinforce other measures, be they military or otherwise? These critical questions can best be answered in comparative case studies.

Another issue concerns the repercussions of rogue nations' behaviour on the potential peer competitors. To assess these, one must combine a study of past cases – historically, how did rogue nations interact with larger, emerging powers? – with an assessment of the current and likely future interests of the potential peer competitors. It may be that rogue nations cause too many problems for these larger powers – they entrap them in undesired conflicts – but that the larger powers nevertheless maintain a strategic interest in countering missile defence. It may also be that this interest is nurtured by US policy towards the rogue nations, causing peers not only to counter missile defence but also to incite conflict between the US and the rogues. Finally, it could be that rogue nations, to the extent that they enter into an 'asymmetric' alliance with terrorists, cause a durable partnership between the US and its potential peer competitors and thus make missile defence recede in importance.

The future of missile defence is thus rogue: rogue nations challenge the ability of the US and allies to cooperate on complex issues, and they influence the level of competition between the US and its potential peers. It is ultimately the US ability to interact with the many regional complexities in these two contexts that will determine whether missile defence reinforces or weakens America's imperial fabric.

Notes

1 'The National Security Strategy of the United States of America', The White House, September 2002, p. 15. Online. Available HTTP: ⟨http://www.whitehouse.gov/nsc/nss.html⟩.

2 See for example ibid., pp. 27–28, or 'Annual Report on the Military Power of the People's Republic of China', Report delivered by the Secretary of Defense to Congress, 28 July 2003, pp. 31–37.

3 James E. Lindsay and Michael E. O'Hanlon, *Defending America: The Case for Limited National Missile Defense*, New York, Brookings, 2001. The authors discuss Russia and China in two contexts: as threats and as part of the international politics of missile defence (see chs 3 and 5 respectively).

INDEX